OBELISKS

TOWERS OF POWER

David Hatcher Childress

Adventures Unlimited Press

Other Books by David Hatcher Childress:

VIMANA
ARK OF GOD
ANCIENT TECHNOLOGY IN PERU & BOLIVIA
THE MYSTERY OF THE OLMECS
PIRATES AND THE LOST TEMPLAR FLEET
TECHNOLOGY OF THE GODS
A HITCHHIKER'S GUIDE TO ARMAGEDDON
LOST CONTINENTS & THE HOLLOW EARTH
ATLANTIS & THE POWER SYSTEM OF THE GODS
THE FANTASTIC INVENTIONS OF NIKOLA TESLA
LOST CITIES OF NORTH & CENTRAL AMERICA
LOST CITIES OF CHINA, CENTRAL ASIA & INDIA
LOST CITIES & ANCIENT MYSTERIES OF AFRICA & ARABIA
LOST CITIES & ANCIENT MYSTERIES OF SOUTH AMERICA
LOST CITIES OF ANCIENT LEMURIA & THE PACIFIC
LOST CITIES OF ATLANTIS, ANCIENT EUROPE & THE MEDITERRANEAN
LOST CITIES & ANCIENT MYSTERIES OF THE SOUTHWEST
YETIS, SASQUATCH AND HAIRY GIANTS

With Brien Foerster
THE ENIGMA OF CRANIAL DEFORMATION

With Steven Mehler
THE CRYSTAL SKULLS

OBELISKS

TOWERS OF POWER

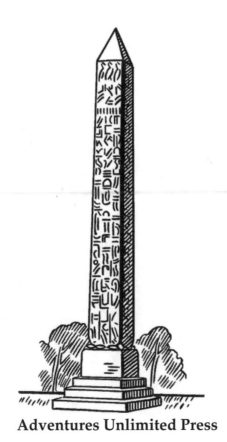

Adventures Unlimited Press

Obelisks: Towers of Power

ISBN 978-1-948803-05-2

Published by:
Adventures Unlimited Press
One Adventure Place
Kempton, Illinois 60946 USA
auphq@frontiernet.net

AdventuresUnlimitedPress.com

10 9 8 7 6 5 4 3 2 1

Two obelisks, still standing in the ruins of Karnak Temple in Luxor, Egypt in the 1920s. Most of the well-built temple was in ruins but the obelisks were still standing. Indeed, they must have been there before the Temple of Karnak was built many thousands of years ago. How the obelisks were erected is still a mystery.

OBELISKS

TOWERS OF POWER

David Hatcher Childress

The Great Seal of the United States with the Great Pyramid and a pyramidion with an eye at its center. What ancient power once projected from the Great Pyramid and the obelisks? This symbol is on the obverse side of the one dollar bill and is part of our everyday lives.

TABLE OF CONTENTS

Chapter 1

The Mysterious Obelisk and the Pyramidion   11

Chapter 2

The Megalith Masterminds   35

Chapter 3

Mysteries of the Unfinished Obelisk   107

Chapter 4

The Problem of the Obelisks   125

Chapter 5

The Obelisks of Ethiopia   177

Chapter 6

Obelisks in Europe and Asia   205

Chapter 7

Obelisks in the Americas   247

Chapter 8

The Towers of Atlantis   263

Chapter 9

Obelisks on the Moon   301

Chapter 10

Bibliography and Footnotes   321

Cleopatra's Needle at Alexandria, Egypt in 1880.

Chapter 1

The Mysterious Obelisk
and the Pyramidion

*Research is a straight line from the tangent
of a well-known assumption to
the center of a foregone conclusion.*
—Neville's Conclusion

For many years I have been fascinated by obelisks and I have included the subject in many of my presentations at conferences around the world. As a collector of books (and other things) I have probably every book on obelisks ever published in English. And that is not very many books—only about six—and they don't have much to say about the reason for the existence of the large monolithic obelisks that were created thousands of years ago. The date of the creation and erection of many obelisks is of some considerable doubt as we shall see as this book progresses.

What will be clear is that obelisks are among the largest and most mysterious of all quarried stone in ancient times. We do not know how many obelisks were quarried or erected in antiquity. We do not know how widespread the erection of obelisks was, nor where the practice originated. We do not know how the largest obelisks were erected and we do not know what function the ancients believed that obelisks served.

Basically, what were obelisks, an engineering feat of Herculean proportions, for? What purpose did they serve? Were they simple monuments to a ray of the sun? Was there some energy that was believed to be transposed and channeled through these gigantic needles of granite? Granite contains tiny resonating quartz crystals infused in the rock structure. Does this matter to the purpose of the obelisk? Are obelisks essentially energy towers of some sort? The answer would seem to be—yes.

First let us discuss some of the early books on obelisks, what few there are, and the standard definitions of the important terms "obelisk" and "pyramidion."

What is an Obelisk?

First of all, let us review the basics of what an obelisk is and how it is defined in the old scientific community. The *Encyclopedia Britannica* defines an obelisk as a:

> ...tapered monolithic pillar, originally erected in pairs at the entrances of ancient Egyptian temples. The Egyptian obelisk was carved from a single piece of stone, usually red granite from the quarries at Aswan. It was designed to be wider at its square or rectangular base than at its pyramidal top, which was often covered with an alloy of gold and silver called electrum. All four sides of the obelisk's shaft are embellished with hieroglyphs that characteristically include religious dedications, usually to the sun god, and commemorations of the rulers. While obelisks are known to have been erected as early as the 4th dynasty (c. 2575–2465 bce), no examples from that era have survived. Obelisks of the 5th dynasty's sun temples were comparatively squat (no more than 10 feet [3.3 meters] tall). The earliest surviving obelisk dates from the reign of Sesostris I (1918–1875 bce) and stands at Heliopolis, a suburb of Cairo, where once stood a temple to Re. One of a pair of obelisks erected at Karnak by Thutmose I (c. 1493–c. 1482 bce) is 80 feet (24 meters) high, square at the base, with sides of 6 feet (1.8 meters), and 143 tons in weight.
>
> An inscription on the base of Hatshepsut's 97-foot (30-meter) standing obelisk at Karnak indicates that the work of cutting that particular monolith out of the quarry took seven months. In the Temple of Hatshepsut at Thebes are scenes of the transport of the obelisk down the Nile by barge. At its destination workmen put the shaft into place upon its detached base by hauling it up a ramp made of earth and tilting it.
>
> Other peoples, including the Phoenicians and the Canaanites, produced obelisks after Egyptian models, although not generally carved from a single block of stone.
>
> During the time of the Roman emperors, many obelisks

were transported from Egypt to what is now Italy. At least a dozen went to the city of Rome itself, including one now in the Piazza San Giovanni in Laterano that was originally erected by Thutmose III (reigned 1479–1426 bce) at Karnak. With a height of 105 feet (32 meters) and a square base with sides of 9 feet (2.7 meters) that tapers to a square top with sides of 6 feet 2 inches (1.88 meters), it weighs approximately 230 tons and is the largest ancient obelisk extant.

Late in the 19th century the government of Egypt divided a pair of obelisks, giving one to the United States and the other to Great Britain. One now stands in Central Park, New York City, and the other on the Thames embankment in London. Although known as Cleopatra's Needles, they have no historical connection with the Egyptian queen. They were dedicated at Heliopolis by Thutmose III and bear inscriptions to him and to Ramses II (reigned c. 1279–c. 1213 bce). Carved from the typical red granite, they stand 69 feet 6 inches (21.2 meters) high, have a rectangular base that is 7 feet 9 inches by 7 feet 8 inches (2.36 meters by 2.33 meters), and weigh 180 tons. The quarrying and erecting of these pillars is a measure of the mechanical genius and the unlimited manpower available to the ancient Egyptians.

A well-known example of a modern obelisk is the Washington Monument, which was completed in Washington, D.C., in 1884. It towers 555 feet (169 meters) and contains an observatory and interior elevator and stairs.

Notice in the article above that it nowhere gives the purpose for obelisks except to say that they "…were erected in pairs at the entrances of ancient Egyptian temples." This is probably the most simple explanation for an obelisk ever given and illustrates that their function—even as seen from a superstitious viewpoint from the ancient Egyptians themselves—is unknown.

This last paragraph is one from which an argument might follow. The Washington Monument—and others like it—is really a faux obelisk because it is not monolithic. In other words, it is built out of thousands of stones and is essentially a huge building with a staircase inside it that looks like an obelisk. A genuine obelisk is a single piece of monolithic stone, mainly granite. And as we shall see, this monolithic needle of granite needs to be perfect and cannot have any cracks or fissures in it.

Similarly, when they say that the Phoenicians built obelisks out of many blocks of stone we cannot be sure just what they are talking about because no obelisks are named. Lebanon has the Temple of the Obelisks, located at Byblos, which contains about 25 small obelisks, the tallest being two and half meters. These structures are small monuments that had a pyramidion—a pyramid shape—on top of them.

The black obelisk of Nimrud.

Wikipedia has a lengthy article on obelisks and mentions obelisks in Assyria. Wikipedia also correctly says that obelisks are monolithic with a pyramidion at the top. Says Wikipedia about obelisks:

> An obelisk is a tall, four-sided, narrow tapering monument which ends in a pyramid-like shape or pyramidion at the top. These were originally called tekhenu by their builders, the Ancient Egyptians. The Greeks who saw them used the Greek term 'obeliskos' to describe them, and this word passed into Latin and ultimately English. Ancient obelisks are monolithic; that is, they consist of a single stone. Most modern obelisks are made of several stones; some, like the Washington Monument, are buildings.
>
> The term stele is generally used for other monumental, upright, inscribed and sculpted stones.
>
> Obelisks were prominent in the architecture of the ancient Egyptians, who placed them in pairs at the entrance of temples. The word "obelisk" as used in English today is of Greek rather than Egyptian origin because Herodotus, the Greek traveler, was one of the first classical writers to describe the objects. A number of ancient Egyptian obelisks are known to have survived, plus the "Unfinished Obelisk" found partly hewn from its quarry at Aswan. These obelisks are now dispersed around the world, and fewer than half of them remain in Egypt.
>
> The earliest temple obelisk still in its original position

is the 68-foot (20.7 m) 120-metric-ton (130-short-ton) red granite Obelisk of Senusret I of the XIIth Dynasty at Al-Matariyyah in modern Heliopolis.

The obelisk symbolized the sun god Ra, and during the brief religious reformation of Akhenaten was said to be a petrified ray of the Aten, the sundisk. It was also thought that the god existed within the structure.

Benben was the mound that arose from the primordial waters Nu upon which the creator god Atum settled in the creation story of the Heliopolitan creation myth form of Ancient Egyptian religion. The Benben stone (also known as a pyramidion) is the top stone of the Egyptian pyramid. It is also related to the Obelisk. It is hypothesized by New York University Egyptologist Patricia Blackwell Gary and Astronomy senior editor Richard Talcott that the shapes of the ancient Egyptian pyramid and obelisk were derived from natural phenomena associated with the sun (the sun-god Ra being the Egyptians' greatest deity).

The pyramid and obelisk might have been inspired by previously overlooked astronomical phenomena connected with sunrise and sunset: the zodiacal light and sun pillars respectively. The Ancient Romans were strongly influenced by the obelisk form, to the extent that there are now more than twice as many obelisks standing in Rome as remain in Egypt. All fell after the Roman period except for the Vatican obelisk and were re-erected in different locations.

The largest standing and tallest Egyptian obelisk is the Lateran Obelisk in the square at the west side of the Lateran Basilica in Rome at 105.6 feet (32.2 m) tall and a weight of 455 metric tons (502 short tons).

Not all the Egyptian obelisks in the Roman Empire were set up at Rome. Herod the Great imitated his Roman patrons and set up a red granite Egyptian obelisk in the hippodrome of his new city Caesarea in northern Judea. This one is about 40 feet (12 m) tall and weighs about 100 metric tons (110 short tons).

It was discovered by archaeologists and has been re-erected at its former site. In Constantinople, the Eastern Emperor Theodosius shipped an obelisk in AD 390 and had it set up in his hippodrome, where it has weathered Crusaders and Seljuks and stands in the Hippodrome square in modern Istanbul. This one stood 95 feet (29 m) tall and weighing 380 metric tons (420 short tons). Its lower half reputedly also once stood in Istanbul but is now lost. The Istanbul obelisk is 65 feet (20 m) tall.

One of the things that is associated with obelisks is the benben stone. What is that?

The Benben Stone

The benben stone is a pyramid-shaped stone that is associated with the primordial beginnings of mankind upon which the creator god Atum settled, as we have seen above. The benben stone was also called a pyramidion by the Greeks. The benben stone and the benben bird are mysterious, but at the least what little we know seems to explain the deep reverence that the ancient Egptians had for the pyramids at Giza. Here is what Wikipedia has to say about the benben:

> In the *Pyramid Texts*, e.g. Utterances 587 and 600, Atum himself is at times referred to as "mound." It was said to have turned into a small pyramid, located in Heliopolis (Egyptian: Annu or Iunu), within which Atum was said to dwell. Other cities developed their own myths of the primeval mound. At Memphis the god Tatenen, an earth god and the origin of "all things in the shape of food and viands, divine offers, all good things" was the personification of the primeval mound.

> The Benben stone, named after the mound, was a sacred stone in the temple of Ra at Heliopolis (Egyptian: Annu or Iunu). It was the location on which the first rays of the sun fell. It is thought to have been the prototype for later obelisks, and the capstones of the great pyramids were based on its design. The capstone or the tip of the pyramid is also called a pyramidion. In ancient Egypt, these were probably gilded, so they shone in sunlight. The pyramidion is also called 'Benben stone.' Many such Benben stones, often carved with images and inscriptions, are found in museums around the world.

> The bird deity Bennu, which was probably the inspiration for the phoenix, was venerated at Heliopolis, where it was said to be living on the Benben stone or on the holy willow

tree.

According to Barry Kemp, the connection between the benben, the phoenix, and the sun may well have been based on alliteration: the rising, weben, of the sun sending its rays towards the benben, on which the benu bird lives. Utterance 600, § 1652 of the *Pyramid Texts* speaks of Atum as you rose up, as the benben, in the Mansion of the Benu in Heliopolis.

From the earliest times, the portrayal of Benben was stylized in two ways; the first was as a pointed, pyramidal form, which was probably the model for pyramids and obelisks. The other form was round-topped; this was probably the origin of Benben as a free standing votive object, and an object of veneration.

During the Fifth Dynasty, the portrayal of benben was formalized as a squat obelisk. Later, during the Middle Kingdom, this became a long, thin obelisk.

In the Amarna Period tomb of Panehesy, the benben is seen as a large, round-topped stela standing on a raised platform.

The Egyptologist Barry Kemp is saying that the Bennu bird is only associated with pyramids and obelisks because it has a similar-sounding name and also flies over the river—or rises up— and the sentence becomes an alliteration. But still, whatever role this Bennu bird has in the mystery of obelisks, we still don't know how the benben stone is related to obelisks. Wikipedia tells us—with no elaboration—the benben is related to obelisks. But how?

Obviously, obelisks typically end with a pointed pyramid at the end tip of their monolithic structure. Egyptian pyramids also typically ended with a point (Hindu, Olmec and Mayan pyramids typically ended with a temple or flat area at the apex) and so we see a similarity with obelisks. But there seems to be some other connection between pyramids, pyramidions, and obelisks. Is it that they are all involved in energy production and transfer?

The benben appears on the reverse side of the dollar bill as the apex to the pyramid. Yes, that eye over the pyramid is the benben stone. It seems curious that everyday Americans—and others—are carrying in their wallets dollar bills that have a benben representation on them, but we don't know what the benben is. Is the benben the "Eye of God" who smiles down upon mankind and specifically the nation known as the United States of America? Rather than being

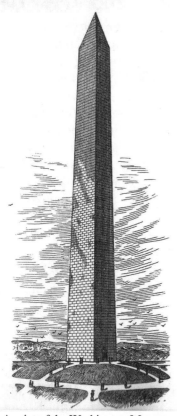

A print of the Washington Monument.

on the tip of an obelisk, this image is on top of the Great Pyramid and it combines the benben stone with the Eye of Horus and the Great Pyramid.

The Power of the Monument

But why is an image of the Great Pyramid on our money? It is part of the Great Seal of the United States and scholars tell us that this is a symbol used by early Masonic lodges in England, Scotland, France and the United States. It is a mystical symbol of the great "Builders" of ancient times, those who built the great temples, pyramids and obelisks of Egypt. The symbol of the benben stone with its Eye of Horus and Great Pyramid below it suggests that an obelisk is also part of this scene. And, indeed, the Washington Monument was built some decades after the Great Seal was designed. It was to symbolize the new power of the United States—and they chose to express that power with a gigantic obelisk-styled building.

In benben talk, the Washington Monument might be described as the fulfillment of an ancient Egyptian prophecy of the return of the powerful and righteous state that was early megalithic Egypt. The early founding fathers and designers of Washington D.C. could not build a structure as mighty and imposing as the Great Pyramid with a golden capstone pyramidion, benben stone or such. But they could build a gigantic obelisk-building—and they did.

Construction of the Washington Monument began in 1848, but the work stopped from 1854 to 1877 due to a lack of funds and the intervention of the Civil War. The internal stone structure was completed in 1884 but the final internal ironwork and other finishing touches were not completed until 1888. Says the official National Park Service website on the Washington Monument: "Built to honor

George Washington, the United States' first president, the 555-foot marble obelisk towers over Washington, D.C."

The Washington Monument is an important symbol of Washington and the power of the United States as a nation of spiritually-oriented people who are dedicated to the rule of divine law, as best they can interpret it. We also see how it is a faux obelisk that is associated with the Great Pyramid, the benben stone and the Eye of Horus.

The Latin mottos on the Great Seal: "He Has Smiled On Our Beginning" and "The New Order of the Ages" suggest a biblical prophecy that is associated with the Great Pyramid as depicted on the Great Seal, featured on the one dollar bill. But does the Bible ever mention the Great Pyramid, let alone obelisks?

While Bible scholars debate the subject of whether the Great Pyramid, or pyramids in general, are mentioned in the Bible, we can say that obelisks are apparently mentioned once, but not

The Great Seal of the United States of America complete with pyramidion.

pyramidions or benben stones.

We do have one reference in Isaiah 19:19 of the Old Testament when the prophet says: "In that day there will be an altar to the Lord in the midst of the land of Egypt, and a pillar to the Lord at its border." This pillar sounds a lot like an obelisk, probably a small one. Perhaps the "altar to the Lord" is the Great Pyramid at Giza, but this is unclear. More on this later.

Cats are not mentioned in the Bible and scholars have noted this. Dogs and various other animals are mentioned, including the Ethiopian ass or donkey, but cats do not appear, even once. Yet, we know that cats existed in Egypt and throughout the Mediterranean and Arabia. Cats were famously worshipped in Egypt and their main function—an important one—was to keep the large state granaries free of mice and other rodents. Their very practical mission of keeping mice from eating the stored wheat of abundant Egypt was greatly appreciated by the populace and the cat was elevated to godly status, usually as Bast or Bastet.

So we might forgive the Bible from barely mentioning obelisks. But does the Bible mention the Great Pyramid?

According to Gotquestions.org the main canonical scriptures, i.e., those in the Bible, do not ever specifically mention the pyramids of Giza. However, the word they would have used would have been "migdol," which is found. Says the site:

> Pyramids are not mentioned as such in the canonical Scriptures. However, the Apocrypha (approved as canonical by Catholics and Coptics) does mention pyramids in 1 Maccabees 13:28-38 in connection with seven pyramids built by Simon Maccabeus as monuments to his parents.
>
> Pre-Alexandrian Jews would not have used the word pyramid. However, in the Old Testament, we do see the word migdol. This word is translated "tower" and could represent any large monolith, obelisk or pyramid. Migdol is the Hebrew word used to describe the Tower of Babel in Genesis 11:4, and it is translated similarly in Ezekiel 29:10 and 30:6. In describing a "pyramid," this is the word the Hebrews would have most likely used. Furthermore, Migdol is a place name in Exodus 14:2, Numbers 33:7, Jeremiah 44:1, and Jeremiah 46:14 and could mean that a tower or monument was located there.

So the ancient word in Egypt and the Middle East for a stele, obelisk, monolith or pyramid was migdol. However, the term is hardly ever used in old texts despite the fact that these monuments were fairly common throughout Egypt and elsewhere. In a similar explanation as to why cats are never mentioned in the Bible, it may be that they were fairly common and not part of any important history to the writers of the texts.

The Bible says that the Israelites were tasked with making mud bricks as slaves in Egypt, and it mentions a place called Migdol in Exodus 14:2:

> Tell the people of Israel to turn back and encamp in front
> of Pi-hahiroth, between Migdol and the sea, in front of Baal-
> zephon; you shall encamp facing it, by the sea.

this is apparently referring to a place with an obelisk or monolith near the sea, probably the Red Sea. Migdol could also describe a pyramid, but it's not likely to be one of the pyramids at Giza because they are not near the sea but on a limestone plateau on the edge of the Western Desert. This area did get flooded every year from the Nile, but it was not until the late 1800s when travel to Egypt became more common and tourism became a major player in economy of the area that groups—literal expeditions at the time—made desert voyages from Cairo to the pyramids in the western sand dunes. The pyramids of Giza were remote and unfamiliar to most Egyptians.

In ancient times it would have been difficult for normal peasants

and merchants to venture into the area of the eastern Nile Delta where the Giza pyramids are located. If an ancient traveler was to have been in the area and witnessed the massive Giza pyramids he would likely have done so in a boat during the annual flooding of the Nile that filled Lake Moeris and turned much of the area into a great lake.

Still, even ancient historians knew about the pyramids and attempted to visit them. The three Pyramids at Giza were the first of the Seven Wonders of the World and the only one of those that still exists and has not been destroyed. Indeed, they are essentially indestructible. Obelisks are tough as well, but they are not indestructible. As we shall see, many obelisks were purposely destroyed in the past or in many cases were toppled and broken in earthquakes or other earth changes.

To revisit Isaiah:19, it is interesting to note that the early part of the chapter describes widespread destruction and devastation that would befall Egypt. Of course, the Israelites see this in terms of their Lord wreaking havoc on their enemy. In fact, the chapter begins with an interesting image:

A prophecy against Egypt: See, the Lord rides on a swift cloud and is coming to Egypt.

The idols of Egypt tremble before him, and the hearts of the Egyptians melt with fear.

Note that the Lord is riding on a "swift cloud" — something that sounds like a flying saucer or UFO.

The succeeding verses detail the woes that will be visited on Egypt. This section speaks of what is possibly an event that happened circa 2000 BC when a dam on the Nile caused flooding in Lower Egypt and the delta region. Some biblical historians think that this event was the failure of the Isis dam across the Nile causing the Sesonchosis (the Season of Chaos) of Egyptian history. Verses 19 and 20 of Isaiah: 19 predict what will happen when the Egyptians capitulate and acknowledge the Israelite Lord:

(19) On that day there shall be an altar to the Lord, In the midst of the Land of Egypt and a pillar at its Border to the Lord. (20) and it shall be a sign and a witness to the Lord of Hosts in the land of Egypt...

This may be referring to the Great Pyramid, located near where the Nile Delta comes together, in some ways the center of Egypt. When it speaks of a "pillar at its border," it may possibly refer to an obelisk. We have noted before that Egyptians placed obelisks in front of their temples. Would this describe a temple to the Lord, built in the Egyptian style?

Despite their importance today obelisks were not discussed in any meaningful way in ancient texts that we are familiar with, and there is a serious lack of source material on these monuments. They therefore don't make it into the discussion and analysis of gigantic stone monuments as pyramids and statues do. Indeed, part of the mystery of obelisks—and pyramids—is that they are hardly ever discussed at all in ancient texts. The reason may be that no one knows anything about them. They exist and are obviously something consequential, but no one knows why obelisks or pyramids were created or how they were erected. Associated with obelisks are pyramidions, often mini-pyramids of one solid piece of stone. What was their purpose?

The Mystery of Pyramidions and Sun Temples

In 1550 BC, more than 1,200 years after the Giza Pyramids were thought to have been erected (they may be older), ancient Egyptian pharaohs were building magnificent temples adorned with huge monuments that pointed to the sky. These were the obelisks, rising toward the heavens and pointing upward with the tips of their pyramid-shaped tops.

As we have seen, an obelisk is a four-sided single piece of stone standing upright, gradually tapering as it rises and terminating in a small pyramid called a pyramidion. Obelisks were known to the ancient Egyptians as Tekhenu, a word whose derivation is unknown. When the Greeks became interested in Egypt, both obelisks and pyramids attracted their attention. To the former they gave the name "obeliskos," from which the modern name in almost all languages is derived. Obeliskos is a Greek diminutive meaning "small spit"; it was applied to obelisks because of their tall, narrow shape. In Arabic, the term is Messalah, which means a large patching needle, and again has reference to the object's form.

Around the Heliopolis area, these monoliths were commonly of red granite from Syene (now known as Aswan) and were dedicated to the sun god. In traditional dynastic Egypt they were usually placed in pairs before the temples, one on either side of the portal.

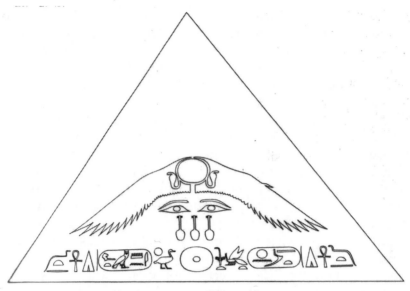

A drawing of a typical pyramidion, often of solid basalt.

Few actual temples with obelisks remain, the main one being Luxor Temple.

Down each of the four faces, in most cases, ran a line of deeply incised hieroglyphs and representations, setting forth the names and titles of the pharaoh. The cap, or pyramidion, was sometimes sheathed with copper or other metal.

According to traditional Egyptology, obelisks of colossal size were first raised in the 12th dynasty. Of those still standing in Egypt, one remains at Heliopolis and two at Karnak, one said to be from the time of Thutmose I.

Queen Hatshepsut is credited with erecting the other obelisk at the Temple of Karnak in Luxor. Queen Hatshepsut (born c. 1482 BC), an Egyptian queen of the 18th dynasty, was the most powerful woman to rule Egypt as a pharaoh. After the death (c.1504 BC) of her husband, Thutmose II, she assumed power, first as regent for his son Thutmosis III, and then (c. 1503 BC) as pharaoh.

One of the theories of this book is that these, like all the famous obelisks, were erected many thousands of years earlier, by another culture. Being virtually indestructible, they later had temples built around them. Originally, we theorize, the obelisks were polished granite without any hieroglyphs. During the reign of succeeding pharaohs in the dynsastic period, they were basically carved with hieroglyphs and dedicated, causing most archeologists to believe

24

the obelisks themselves were made when they were dedicated.

According to American University of Cairo professor Labib Habachi, obelisks are the most often seen—and best known—of all the objects and structures of the ancient civilizations of the past, including all things Mesopotamian, Greek, Roman, or Egyptian.

Professor Habachi is the author of *The Obelisks of Egypt*[1] (1984, American University Press, Cairo) that is one of the few books published on the subject. The other significant books are the 1923 tome *The Problem of the Obelisks*[8] (which we will discuss shortly in chapter four), *The Magic of Obelisks*[2] by Peter Tompkins (1981), and the more recent book out of Cambridge, Massachusetts, *Obelisk: A History*,[7] published in 2000.

Says Professor Habachi (who died in 1984) on the familiarity of obelisks around the world:

> Some of the smaller obelisks and fragments of larger ones are familiar to the numerous visitors of museums in various countries; larger ones which are still on their original sites are admired by the thousands of people who visit Egypt each year. Still others are seen by the crowds who pass through London, Paris, New York, Istanbul, and especially Rome, where there are more obelisks than in any other place.[1]

Habachi says that obelisks were considered by the ancient Egyptians to be sacred to the sun god, whose main center of worship was at Heliopolis, the ruins of which lie in the district of Matariya near Cairo. Although the well-known obelisks date from the 20th century BC, such monuments seem to have been erected there in honor of the sun god in much earlier periods.

According to Habachi, a type of stone resembling the pyramidion of an obelisk was apparently considered sacred to the sun god even before the appearance of the first pharaoh

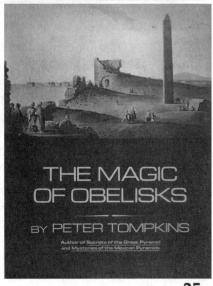

THE MAGIC OF OBELISKS

BY PETER TOMPKINS

Author of Secrets of the Great Pyramid and Mysteries of the Mexican Pyramids

in the First Dynasty (c. 3100-2890 BC). Such stones, which, as we have seen, are known as ben or benben, were believed to have existed in Heliopolis from time immemorial and were the fetish of the primeval god Atum (the setting sun) and the god Re or Re-Harakhti (the rising sun). Habachi is essentially saying that the use of obelisks is "predynastic." I theorize that all of the large obelisks are predynastic. There is no recognized scientific technique for dating when a granite rock was quarried as yet.

Peter Tompkins circa 1980.

As mentioned above, Habachi says that benben stones were associated with obelisks and with the Benu bird, or phoenix. This creature, which begot itself, was thought to have come from the east to live in Heliopolis for 500 years and then to return to the east to be buried by the young phoenix that would in turn replace it in Heliopolis. According to one version of the tale, instead of being replaced the bird revived itself in a burst of flames, and thus it was connected with the god of the dead. In some tombs, an image of the phoenix is shown among the gods. Is this Benu bird some sort of allegory for the obelisk lighting up or glowing?

In the pyramids of the last king of the Fifth Dynasty and the kings of the Sixth Dynasty (c. 2345-2181 BC), the walls of the burial chamber were decorated with Pyramid Texts, religious texts concerned with the welfare of the deceased. One text reads: "O Atum, the Creator. You became high on the height, you rose up as the benben-stone in the mansion of the 'Phoenix' in Heliopolis."

Habachi says that Pliny the Elder (AD 23-79), the Roman encyclopedist, wrote that obelisks were meant to resemble the rays of the sun. This comparison finds support in an inscription addressed to the sun god: "Ubenek em Benben" ("You shine in the benben stone.")[1]

This is a curious inscription because it implies that somehow the obelisk is shining or lit up in the same way as the sun or a ray

26

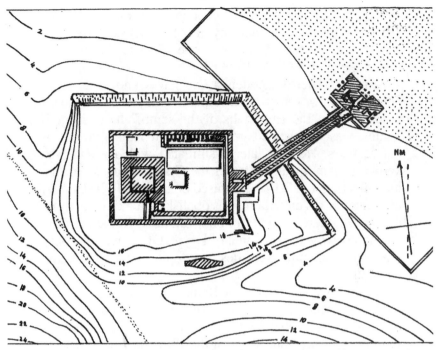

A drawing of an Egyptian Sun Temple from Habachi.

of the sun. How was it that an obelisk had the sun shining inside the stone—was the tip of the obelisk (the benben stone) somehow glowing or emitting energy?

During the prosperous days of the 18th Dynasty (1570-1320 BC), and perhaps at other times, Habachi says the pyramidions of obelisks were covered with gold or some other metal. Habachi says the date at which obelisks were first erected is not known, but the kings of the Fifth Dynasty (2494-2345 BC), who were fervent worshipers of the sun god, may have been the earliest rulers to decorate the facades of their temples with pairs of such monuments.

Says Habachi about the places obelisks were erected:

> Heliopolis, the city of the sun, was called by the ancient Egyptians Iunu, a name meaning "the pillar," and sometimes Iunu Meht, "the northern pillar." The name Iunu appears in the Bible as On; Heliopolis is the Greek name by which the city is generally known. Heliopolis was sacred to the sun god Re and his ennead, a group of nine associated gods. Other gods worshiped there included Kheperi, the scarab, and Shu, the god of the air. Obelisks were first erected at

Heliopolis and the practice was continued throughout the pharaonic period. The majority of these obelisks have been removed or destroyed; the only one still standing there is that of Sesostris I (1971-1928 BC).

Ancient Thebes (modem Luxor) was known as Uast, "the scepter," or sometimes as Iunu Shemayit, "the southern pillar," or as "the Heliopolis of the south." Its main god, Amun, was represented in human form with a crown of tall feathers. He was later assimilated with Re and was known as Amun-Re, "King of the Gods." Because of this identification, obelisks were raised on his behalf. In Thebes, the center of his cult, numerous obelisks, including many of the largest, were erected in honor of Amun Re, at the time when the city was the capital of Egypt. Of its obelisks, only three survive; some were destroyed and a few were taken abroad.

Piramesse—that is, Per-Ramessu, "the domain of Rameses" became the capital of Egypt in the reign of Ramesses II (1304-1237 BC) and remained so under the succeeding Ramesside kings of the 19th and 20th dynasties (1320-1200 and 1200-1085 BC). It was embellished with a score of obelisks, for the most part fashioned by Ramesses II, although several made by earlier kings were taken over by him. Most of these obelisks were smaller than those of Thebes. The cults of the great gods, Re, Amun-Re, and Ptah of Memphis, were introduced in the new capital, and the names of these and other gods appear upon the obelisks Ramesses II erected there.[1]

Elsewhere, according to Habachi, only rather small obelisks have been found. The inscriptions on these make it clear that they were erected in honor of local divinities who were either solar gods or associated with the solar cult. Two pairs of obelisks which were recovered from the ruins on the island of Elephantine near Aswan were dedicated to Khnum, the ram-headed god who fashioned mankind upon a potter's wheel. He was later associated with the sun god Re and was known as Khnum-Re. His obelisks say they were set up at the "altar of Re"—probably in a solar chapel. During the nineteenth century a pair of Ptolemaic obelisks dedicated to Isis was unearthed on the island of Philae. On them are mentioned the solar gods Atum and Amun Re. At Abu Simbel, a chapel of Re Harakhti

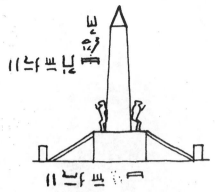

adjacent to the Great Temple of Ramesses II, contained a pair of obelisks and other cult objects related to the sun god.

According to Habachi, an obelisk found at Minshah in Middle Egypt undoubtedly once stood at the neighboring religious center of Abydos. On this obelisk the king is called "beloved of Osiris," the

A depicition of baboons worshipping an obelisk.

god of the dead, who was the principal god of Abydos, although other deities also had cult places there. He says that there were two obelisks in Ashmunein, also in Middle Egypt, both dedicated to Thoth, god of writing and wisdom, and titulary deity of the place.

Among Thoth's many attributes was that of "representative of Re." The ibis and the baboon were sacred to Thoth, and the latter animals were often shown adoring the sun god. In the quarries of Gebel el-Ahmar near Cairo an obelisk is depicted standing between two baboons with their front legs raised in worship. The curious depiction of the Thoth baboons worshipping an obelisk is strange, indeed. Did some sort of power or energy come out of the obelisk?

Says Habachi:

> An inscription on a fragment of an obelisk from Horbeit in the eastern part of the Delta mentions Osiris and his sacred bull, the Mnevis, known as "the living soul of Re." At Athribis in the center of the Delta the pedestals and a few fragments of the shafts of two obelisks still remain. On one fragment, the king is shown with local divinities, one of which is Atum. A number of obelisks were also raised in honor of Atum, a local god of Sais, which was a political center in the Delta and capital of the country during the Saite Period (664-525 BC).
>
> The kings who erected obelisks were usually described on them as beloved of various local and solar gods, and in many cases the king was shown in close relationship to these divinities. One text from an obelisk describes the king as "appearing like Harakhti, beautiful as King of the Two

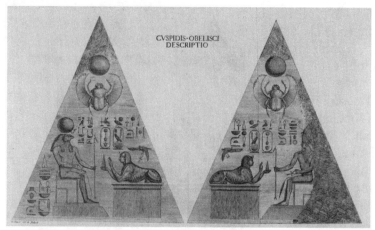

An engraving from 1750 of the pyramidion of the Montecitorio Obelisk in Rome.

Lands like Atum," and a second, as "the one whom Atum made to be King of the Two Lands and to whom [he] gave Egypt, the desert, and foreign lands."

On some obelisks there are references to royal victories, but these are rarely actual historical events. The boasts on most of the obelisks erected by Ramesses II are particularly suspect. On one, this king is commemorated as "the one who defeats the land of Asia, who vanquishes the Nine Bows, who makes the foreign lands as if they were not." On another it is said, "His power is like that of Monthu [the god of war], the bull who tramples the foreign lands and kills the rebels." The king is described as recipient of tribute, noble governor, brave, and vigilant.

If the claims of Ramesses II are not justified, those of his predecessor Tuthmosis III carry greater weight. In celebration of the great victory which Tuthmosis III gained over his powerful enemies in Asia, he erected two obelisks at the Temple of Karnak; the upper part of one of these survives in Istanbul. On one side of it, the king is spoken of as "the lord of victory, who subdues every [land] and who establishes his frontier at the beginning of the earth [the extreme south] and at the marshland up to Naharina [in the north]." On another side, he is said to have crossed the Euphrates with his army to make great slaughter. This crossing of the river was a great achievement, equaled only by his grandfather Tuthmosis I. It provided sufficient justification for the erection of the obelisks.

Yet another reason for setting up obelisks is indicated on the Istanbul obelisk, where Tuthmosis III is described as "a king who conquers all the lands, long of life and lord of Jubilees." Beginning in the thirtieth year of a king's reign, and every three years thereafter, a festival of renewal was celebrated. On the occasion of these jubilees, the kings set up obelisks. The obelisk of Queen Hatshepsut (1503-1482 BC), which still stands at the Temple of Karnak, describes her as "the one for whom her father Amun established the name 'Makare' upon the Ashed-tree [a tree of eternity] in reward for this hard, beautiful, and excellent monument which she made for her First jubilee." However, since Hatshepsut reigned only about twenty years, she evidently celebrated her jubilee much earlier.[1]

Habachi says that obelisks were always regarded by the ancient Egyptians as a symbol of the sun god related to the Benben, but during certain periods they were looked upon as being themselves occupied by a god and thus entitled to offerings. He says that this is the case with four obelisks erected by Tuthmosis III in the Temple of Amun-Re at Karnak. In an inscription there, the king recorded the establishment of new feasts and offerings which he instituted for the four obelisks, dedicating 25 loaves of bread and jars of beer to

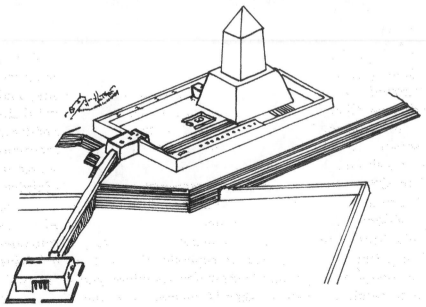

Habachi's drawing of an Egyptian Sun Temple with its short obelisk and buildings.

31

each of them daily. When an obelisk was erected, scarabs showing the king kneeling in adoration before it were issued. A vignette accompanying the 15th chapter of the *Book of the Dead,* a guide for the dead in their travels through the Underworld, is entitled "Adoring Re-Harakhti when he rises in the eastern horizon of the sky." In the scene are two priests, one reciting from a roll of papyrus which he holds, the other making offerings to two obelisks which embody Re-Harakhti.

In addition to large obelisks, smaller obelisks—or rather obeliskoid objects—were sometimes placed in front of tombs. These objects were inscribed on only one face with the name and the main title of the tomb owner. Some bear prayers addressed to the gods of the dead on behalf of the tomb owner. During the Old Kingdom, the kings erected pyramids to serve as their burial places. In fact, the pyramids were but a focus of a large funerary complex with a mortuary temple at the base of the pyramid connected by means of a causeway to a valley temple on the edge of the cultivated area. The kings of the Fifth Dynasty, already mentioned as especially devoted to the sun god, added to their pyramid complexes solar temples in which a gigantic obelisk was the main feature.

These strange "solar temples" with their obelisks and causeways look a lot like a ceremonial imitations of an actual power station. What sort of rituals were the dynastic Egyptians using to worship these ancient towers? Why were obelisks thought to have some sort of divine power? How were they even quarried and erected? And why would anyone go through such tremendous effort to raise a seemingly non-functional monument?

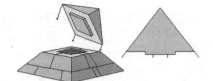

Diagram of a pyramidion block.

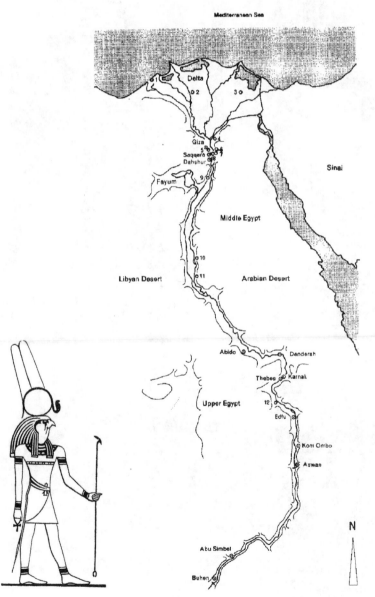

A map of ancient Egypt including Giza and Aswan.

34   An old print of angels helping to erect one of the obelisks brought to Rome.

Chapter 2

The Megalith Masterminds

Facts do not cease to exist because they are ignored.
—Aldous Huxley

Truth is one, but error proliferates.
Man tracks it down and cuts it up into
little pieces hoping to turn it into grains of truth.
—René Dauman, *The Way of the Truth*

Megalithomania

Legends of resplendent ancient civilizations and their cataclysmic destruction are part of nearly every culture in the world. The modern skeptic asks, "Well, if highly advanced ancient societies existed in the past then where is the evidence of their machinery and such; and shouldn't there be the remains of these peoples' cities?" The answer is that such evidence does exist, and hundreds of ruined cities have been found both above and below water.

The idea that man was primitive in the past and that the present represents the highest civilization our planet has achieved is fairly well accepted in the West, while other cultures view history as cyclical and see our current society as a decline from a former golden age. Out of the past we find megalithic cities built to last for thousands of years. How primitive must we suppose these people were?

Throughout the world, there exists a type of megalithic construction that is called "Atlantean" by those researchers who believe in advanced civilizations of the past. This is typically a type of construction that used gigantic blocks of stone, often crystalline granite (granite is infused with tiny quartz crystals). Huge blocks may be fitted together without mortar in a polygonal style that tends to interlock the heavy blocks in a jigsaw fashion. These interlocked polygonal walls resist earthquake damage by moving with the shock wave of the quake. They momentarily jumble themselves and move freely but then fall back into place. These interlocked jigsaw walls will not collapse with an earthquake shock wave as with brick wall construction.

Such "Atlantean style" construction can be found all over the world, and certain large structures are termed "Atlantean" even by mainstream archeologists. Classical examples of such construction are at Mycenae in the Greek Peloponnese and the temples of Malta, along with the nauraghi of Sardinia and Stonehenge and Carnac in Europe. The gigantic megalithic walls of Tiahuanaco, Sacsayhuaman and Ollantaytambo in South America, Monte Alban in Mexico as well as the pre-Egyptian structures of the Osirion at Abydos and the Valley Temple of the Sphinx can all be counted Atlantean works. Other megaliths, including obelisks, can be found at the Ethiopian site of Axum, which will be discussed in another chapter.

Atlantean architecture is often circular, and uses the most exact rock cutting techniques to fit blocks together. Atlantean-type architecture often uses "keystone cuts"—identical shapes are cut into rock on both sides of a joint and the space is fitted with a metal clamp. These keystone cuts are typically an hourglass shape or a double-T shape. The clamps that went inside them may have been copper, bronze, silver, electrum (a mixture of silver and gold) or some other metal. In nearly every case where keystone cuts can be found, the metal clamp has already been removed—many thousands of years ago!

Obelisks easily fit into this style of construction and all of the problems of quarrying, removing, transporting and erecting megaliths that weigh many tons are multiplied many times over because of the sheer size and weight of the gigantic stone. We might easily assume that if there had been a civilization known as Atlantis they would have quarried large granite blocks, including obelisks.

Unlike the ancient Egyptians, who were said to have only copper chisels and saws—and not iron ones—the Atlanteans (who ever they were) could be said to have iron tools, diamond saws, cranes and pulleys, and even airships and anti-gravity technology to easily move the gigantic stones that weigh hundreds of tons. Why else would anyone go through the many difficulties of moving and erecting gigantic blocks of granite if it was not something that they could do with ease, given the amazing technology that they had. Indeed, in theory this technology is not unlike the technology that we possess today.

But modern Egyptologists tell us that these obelisks and other megaliths were raised in antiquity by the dynastic Egyptians who had only copper tools and did not know about pulleys or what is known as a block and tackle. They did have levers, scaffolding and stone hammers. Yet, were obelisks created and erected using only primitive techniques? Why would anyone have wanted a 500-ton monolith anyway? And, especially, a 500-ton monolith that appears to serve no function? Are some obelisks from an earlier age such as that of Atlantis?

Many well-known, and not-so-well-known, ruins of the world contain the remains of even earlier cities within them. Such sites as Baalbek in Lebanon, Cuzco in Peru, the Acropolis of Athens, Lixus of Morocco, Cadiz in Spain and even the Temple Mount of Jerusalem are built on gigantic remains of earlier ruins. Some modern cities, Cuzco is a good example, contain three or more levels of occupation, including modern occupants. Some archaeologists think the civilization that these earlier buildings are from is the "mythical" civilization of Atlantis.

The idea that man has only recently invented such things as electricity, generators, steam and combustion engines, or even powered flight does not necessarily hold true for a world that rides the rollercoaster of history.

Indeed, when we see how quickly inventions are absorbed into today's society we can imagine how quickly a highly scientific civilization may have arisen in remote antiquity. Just as today there are still primitive tribes in New Guinea, India and South America who live a Stone Age existence, so could Atlantis have existed during a period where other areas of the world lived in various states of development.

The ancient world of Atlantis may have been a lot like the modern world of today—juxtaposed between various factions in the government and military, while international discontent rises in various client colonies of an economic system set up by large business interests. According to the mythos that has built up around Atlantis, it was destroyed because of the wars that it had fought around the world. Today the world is again teetering on the brink of wholesale Armageddon because of political, religious and ethnic differences. Does modern man have something to gain by studying the past? Students of Atlantis believe so.

Osiris.

The Osirian Civilization

The Osirian Civilization, according to esoteric tradition, was an advanced civilization contemporary with Atlantis. In the world of about 15,000 years ago, there were a number of the highly developed and sophisticated civilizations on our planet, each said to have a high degree of technology. Among these fabled civilizations was Atlantis, while another highly developed civilization existed in India. This civilization is often called the Rama Empire.

What is theorized is a past quite different from that which we have learned in school. It is a past with magnificent cities, ancient roads and trade routes, busy ports and adventurous traders and mariners. Much of the ancient world was civilized, and such areas of the world as ancient India, China, Peru, Mexico and Osiris were thriving commercial centers with many important cities. Many of these cities are permanently lost forever, but others have been or will be discovered!

It is said that at the time of Atlantis and Rama, the Mediterranean was a large and fertile valley, rather than a sea as it is today. The Nile River came out of Africa, as it does today, and was called the River Styx. However, instead of flowing into the Mediterranean Sea at the Nile Delta in northern Egypt, it continued into the valley, and then turned westward to flow into a series of lakes, to the south of Crete. The river flowed out between Malta and Sicily, south of Sardinia

38

and then into the Atlantic at Gibraltar (the Pillars of Hercules). This huge, fertile valley, along with the Sahara (then a vast fertile plain), was known in ancient times as the Osirian civilization.

The Osirian civilization could also be called "Predynastic Egypt," the ancient Egypt that built the Sphinx and pre-Egyptian megaliths such as the Osirion at Abydos. In this outline of ancient history, it was the Osirian Empire that was invaded by Atlantis, and devastating wars raged throughout the world toward the end of the period of Atlantis' warlike imperial expansion.

Solon relates in Plato's dialogues that Atlantis, just near the cataclysmic end, invaded ancient Greece. This ancient Greece was one that the people we call the "ancient" Greeks knew nothing about. This "unknown ancient Greece," we shall see, is closely connected to the Osiris-Isis civilization.

The story of Osiris himself, as related by the Greek historian Plutarch, is revealing in technology. According to Egyptian mythology Osiris was born of the Earth and Sky, was the first king of Egypt and the instrument of its civilization. Osiris allegedly traveled throughout the world, teaching the arts of civilization after the flood. He first weaned the Osiris people of their barbarous ways, taught agriculture, formulated laws and taught the worship of the gods. Having accomplished this, he set off to impart his knowledge to the rest of the world.

During his absence, his wife Isis ruled, but Osiris' brother and her brother-in-law, Typhon (also known as Set, and known to us as Satan) was always ready to disrupt her work. When Osiris returned from civilizing the world (or attempting to, at least), Set/Typhon/Satan decided he would kill Osiris and take Isis for himself. He collected 72 conspirators to his plot and had a beautiful chest made to the exact measurements of Osiris. He threw a banquet and declared that he would give the chest to whoever could lie comfortably within it. When Osiris got in, the conspirators rushed to the chest and fastened the lid with

Isis with Osiris and Horus.

nails. They then poured lead over the box and dumped it into the river where it was carried out to into the Mediterranean, which was much smaller at the time. When Isis heard of Osiris' death, she immediately set out to find her beloved.

The box with Osiris in it came aground at Byblos in present day Lebanon, not too far from the massive slabs at Baalbek. A tree grew around where the box landed, and the king of Byblos had it cut down and used it as a pillar in his palace, Osiris still being inside. Isis eventually located Osiris and brought him back to Egypt, where Typhon (Set/Satan) broke into the box, chopped Osiris into 14 different pieces, and scattered him about the countryside.

The loving Isis went looking for the pieces of her husband, and each time she found a piece she buried it—which is why there are temples dedicated to Osiris all over Egypt. In another version, she only pretends to bury the pieces, in an attempt to fool Set/Typhon, and puts Osiris back together, bringing him back to life. Eventually she found all the pieces, except the phallus, and Osiris, one way or another, returned from the underworld and encouraged his son Horus (the familiar hawk-headed god) to avenge his death. Scenes in Egyptian temples frequently depict the hawk-headed Horus spearing a great serpent, Typhon or Set, in a scene that is identical to that of St. George and the dragon, though depicted thousands of years earlier.

In the happy ending, Isis and Osiris get back together, and have another child, Harpocrates. However, he is born prematurely and is lame in the lower legs as a result.[48]

There are many important themes in the legend of Osiris, including resurrection and the vanquishing of evil by good, and perhaps a key to the ancient Osirian civilization. Were the 14 scattered pieces of Osiris an allusion to 14 sacred sites built by the Osirians throughout the Mediterranean? I have already mentioned the theory that the Mediterranean was once a fertile valley with many cities, farms and temples. Perhaps some of the 14 sites lie still undetected underwater and others are known but their full importance has not been identified. I believe that the early megalithic construction at Baalbek, Jerusalem, Giza and the Osirion at Abydos would count as known sites among the number.

A key to the megalithic society of Osiris can be found in the curious buried ruins of the Osirion (the megalithic, pre-dynastic ruins at Abydos in southern Egypt). The British archeologist Naville noted in a *London Illustrated News* article in 1914 that "here and there on the huge granite blocks was a thick knob... which was used for moving the stones. The blocks are very large—a length of fifteen feet is by no means rare; and the whole structure has decidedly the character of what in Greece is called cyclopean. An Egyptian example of which is at Ghizeh, the so-called temple of the Sphinx."

Naville is directly relating the Osirion to the gigantic and prehistoric construction in Greece and also to the temple of the Sphinx. Other such sites around the former Osirian Empire are on the islands of Malta and Sardinia, in Lebanon and Israel, the Balearic Islands, and other areas. Furthermore, the knobs which may or may

A drawing of the Osirion at Abydos by The *London Illustrated News*.

not be for moving the stones, are the same sort of knobs that occur on the gigantic stones that are used in the massive walls to be found in the vicinity of Cuzco, Peru.

The lack of inscriptions in the Osirion building indicates that the building, like the Valley Temple of the Sphinx, was built before the use of hieroglyphics in Egypt! We know this because the Egyptians always engraved hieroglyphics and decorations into any of their architecture. The only known buildings without such markings, such as the Great Pyramid, the Osirion, and the Valley Temple of the Sphinx, are thought by many archeologists now to be older than other structures. The Osirion is evidently a relic from the civilization of Osiris itself.

The present and the past are
perhaps both present in the future
and the future is contained in the past.
—T.S. Eliot

Baalbek and Osiris

One of the most astonishing ancient ruins in the world is the megalithic base of Baalbek, the pre-Roman ruins upon which a Roman-era temple sits.

The archaeological site of Baalbek is 44 miles east of Beirut and consists of a number of ruins, a quarry and a series of catacombs. Here we find gigantic blocks of stone that weigh up 1000 tons or more. Some of these gigantic stones are still at the nearby quarry while others are perfects fitted into the lower parts of the massive temples to Ba'al and Astarte. These are the largest known quarried blocks of stone, though the largest of the obelisks come very close to their astounding weights. In 2014 it was announced at Baalbek that the largest block of all had been uncovered from under rubble at that quarry and that its weight was calculated at an astonishing 1,650 tons. This makes it the largest quarried stone that has so far been discovered. The Unfinished Obelisk at Aswan, to be discussed in another chapter, would have weighed an estimated 1,168 tons if it had been extracted. Both of are incredible size and would challenge any modern construction company. Were they ultimately moved by some sort of levitation or anti-gravity? It seems like the easiest solution.

The temples to Ba'al and Astarte may initially have been built as part of a prehistoric Sun Temple, and even then on the ruins of

An early photo of one of the giant stones at the quarry at Baalbek in Lebanon, c. 1880.

the more ancient structure, its purpose unknown. The Greeks called the temple "Heliopolis" which means "Sun Temple" or "Sun City." Even so, the original purpose of the gigantic platform may have been something else entirely.

Baalbek is a good example of what happens to large, well-made ancient walls—they are used again and again by other builders who erect a new city or temple on top of the older one, using the handy stones that are already to be found at the site. Often, the original

A drawing of the theoretical use of pulleys by the Romans to build Baalbek.

stones are so colossal that they could not be moved and placed elsewhere anyway. This is exactly what has been going on at many sites, in the Old World and in the Americas. Examples of very ancient stonework (3,000 to 6,000 years old)

A photo of the 1,650-ton block found at Baalbek.

mixed with more recent ancient stonework (500 to 2,500 years ago) can be seen at Monte Alban in Mexico and at such Andean sites as Chavin, Cuzco and Ollantaytambo.

At Baalbek, the Roman architecture (largely destroyed by an earthquake in 1759) does not pose any archaeological problems, but the massive cut stone blocks beneath it certainly do. One part of the enclosure wall, called the Trilithon, is composed of three blocks of hewn stone that are the largest stone blocks ever used in construction on this planet, so far as is known (underwater ruins may reveal larger constructions). This is an engineering feat that has never been equaled in history.

The weight and even size of the stones is open to controversy. According to the author Rene Noorbergen in his fascinating book *Secrets of the Lost Races*,[38] the individual stones are 82 feet long and 15 feet thick and are estimated to weigh between 1,200 and 1,500 tons each (a ton is 2,000 pounds, which would make the blocks weigh an estimated 2,400,000 to 3,000,000 pounds each). While Noorbergen's size may be incorrect, his weight is probably closer to the truth. Even conservative estimates say that the stones weigh at least 750 tons each, which would be one and a half million pounds.[38]

It is an amazing feat of construction, for the blocks have been raised more than 20 feet in order to lie on top of smaller blocks. The colossal stones are fitted together perfectly, and not even a knife blade can be fitted between them.[38] Even the blocks on the level below the Trilithons are incredibly heavy. At 13 feet in length, they probably weigh about 50 tons each, an extremely large-sized bunch of stones by any other estimate, except when compared to the Trilithons. Yet, even the Trilithons are not the largest of the stones!

What was until recently considered the largest hewn block (until the recent 2014 discovery), is 13 feet by 14 feet and nearly 70 feet long and weighing at least 1,000 tons (both Noorbergen and Charles Berlitz[44] give the weight of this stone at 2,000 tons[38]), lies in the nearby quarry which is half a mile away. 1,000 tons is an incredible two million pounds! The stone, located at the quarry, is called "Stone of the South" or Hadjar el Gouble, Arabic for "Stone of the Pregnant Woman."

Noorbergen is correct in saying that there is no crane in the world that could lift any of these stones, no matter what their actual weight is. The largest cranes in the world are stationary cranes constructed at dams to lift huge concrete blocks into place. They can typically lift weights up to several hundred tons. 1,000 tons, and God forbid, 2,000 tons, are far beyond their capacity. How these blocks were moved and raised into position is beyond the comprehension of engineers.

Large numbers of pilgrims came from Mesopotamia as well as the Nile Valley to the Temple of Ba'al–Astarte. The site is mentioned in the Bible in the Book of Kings. There is a vast underground network of passages beneath the acropolis, which were possibly used to shelter pilgrims, probably at a later period. Or, were they for something else? Perhaps they led to undiscovered passages and rooms deeper into the earth.

Who built the massive platform of Baalbek? How did they do it? According to ancient Arab writings, the first Ba'al–Astarte temple, including the massive stone blocks, was built a short time after the Flood, at the order of the legendary King Nimrod, by a "tribe of giants."[38]

Nimrod is apparently a king in the area of Baalbek and Syria and he is mentioned in Genesis 10:8–12 as "the first on earth to be a mighty man. Genesis 10:8–12 is known as the Table of Nations which is the lineage of Noah, the famous survivor of the flood, and it says that Nimrod is the great-grandson of Noah, whose son Ham was the father of Cush. Nimrod was "a mighty one in the earth." Nimrod is said to be the son of Cush, but the land of Cush is Ethiopia, Sudan and most of the Red Sea coastline on both sides of this narrow waterway. Nimrod, the son of Cush, is the ruler of Assyria and Baalbek. Says the *Encyclopedia Britannica* about

Nimrod:

...The only other references to Nimrod in the Bible are Micah 5:6, where Assyria is called the land of Nimrod, and I Chronicles 1:10, which reiterates his might. The beginning of his kingdom is said in the Genesis passage to be Babel, Erech, and Akkad in the land of Shinar. Nimrod is said to have then built Nineveh, Calah (modern Nimrud), Rehoboth-Ir, and Resen.

There is some consensus among biblical scholars that the mention of Nimrod in Genesis is a reference not to an individual but to an ancient people in Mesopotamia. The description of Nimrod as a "mighty hunter before the Lord" is an intrusion in this context, but probably, like the historical notices, derived from some old Babylonian saga. However, no equivalent of the name has yet been found in the Babylonian or other cuneiform records. In character there is a certain resemblance between Nimrod and the

An early photo of one of the giant stones at the quarry at Baalbek, c. 1900.

An early photo of some of the gigantic stones at Baalbek.

Mesopotamian epic hero Gilgamesh.

Ancient astronaut theorists have frequently suggested that Baalbek was built by extraterrestrials. Charles Berlitz says that a Soviet scientist named Dr. Agrest suggests that the stones were originally part of a landing and takeoff platform for extraterrestrial spacecraft.[44] The late author and Sumerian scholar Zechariah Sitchin believes, in a like manner, that Baalbek is a launching pad for rockets. He says that extraterrestrials were coming to earth in rockets and used Baalbek as a launch pad when attempting to return their own planet.

While ancient astronauts may well have visited earth in the past, it seems unlikely that they would have arrived here in rockets. They would have mastered the art of more sophisticated spacecraft technologies like anti-gravity and field propulsion and their spaceships would be electric solid-state models, at the very least. One would think such craft would be able to teleport and achieve hyperspace. These craft are submarines as well, if modern UFO phenomena are any indicator of extraterrestrial technology. Such aircraft could land and take off in a pleasant grassy field, and would not need a gigantic platform as the base for a mighty rocket blast.

What then was Baalbek and who built it? The theory of Baalbek being some remnant of the Osirian Empire, along with some of the other megalithic sites in the Mediterranean, fits in well with the Arab legend mentioned previously: that the massive stone blocks were built a short time after the Flood, at the order of King Nimrod.[64]

In his book *Baalbek*[39] archaeologist Friedrich Ragette attempts to explain how Baalbek was built and how the stones were moved into place. Explaining Baalbek is no easy task, Ragette admits, but he does his best.

Ragette first explains that there are two quarries, one about two kilometers north of Baalbek and a closer quarry where the largest stone block in the world still lies. He then makes this interesting remark about the quarries:

> After the block was separated on its vertical side, a groove was cut along its outer base and the piece was felled like a tree on to a layer of earth by means of wedging action from behind. It seems that the Romans also employed a sort of quarrying machine. This we can deduce from the pattern of concentric circular blows shown on some blocks. They are bigger than any man could have produced manually, and we can assume that the cutting tool was fixed to an adjustable lever which would hit the block with great force. Swinging radii of up to 4 m (13 ft) have been observed.[39]

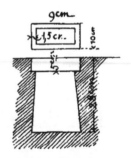

Ragette goes on to theorize that moving an 800-ton stone on rollers would be possible:

> [I]f we assume that the block rested on neatly cut cylindrical timber rollers of 30 cm (12 in) diameter at half-meter distances,

Ragette's keystone wedge.

49

each roller would carry 20 tons. If the contact surface of the roller with the ground were 10 cm (4 in) wide, the pressure would be 5 kg/cm^2 (71 lbs/in^2), which requires a solid stone paving on the ramp. The theoretical force necessary to move that block horizontally would be 80 tons. Another possibility is that the whole block was encased in a cylindrical wrapping of timber and iron braces.

Ragette dismisses this second idea as unlikely and cumbersome. "Also there remains the question of how the block would have been unwrapped and put in place, which brings us to the even more perplexing problem of lifting great weights."[39]

There is not a contractor today that would attempt to move or lift

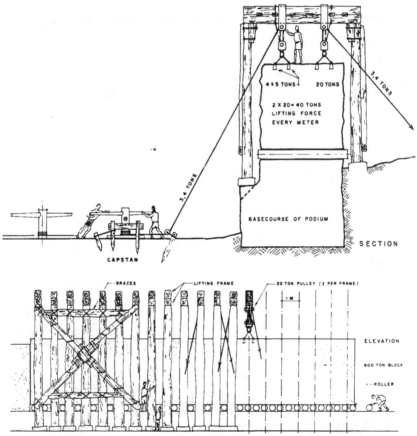

The huge cage proposed in order to lift the massive stones used to build Baalbek.

these stones. It is simply beyond our modern machine technology. I find it interesting that there is no discernible road between the quarry and the massive Sun Temple. This indicates one or both of two possibilities: the building of the lower platform occurred at such an ancient time in antiquity that the road is long gone, or a road was never needed for transporting the blocks. As the INFO article points out, a road would have been of little use anyway.

Ragette cannot solve the problem of lifting such a block into place, saying that it is impossible to lift a huge block such as this completely off the ground by the use of levers. He says that we know that the stone had to be lifted so that the log rollers could be removed from underneath the block and then the block lowered into place. In order to fit perfectly, the stone probably had to be lifted and lowered into place several times at least.

His suggestion is that a giant lifting frame was built around the block and then at least 160 "Lewis" stones—wedge-shaped keystones with metal loops—were inserted into the top of the block. Then a system of pulleys and tackles were used with thousands of manual workers to raise and lower the gigantic blocks a few inches.

Ragette makes no suggestion as to why the Romans, or anyone else, would go to such immense trouble, attempting a virtually impossible engineering feat, of moving these stone into place. Cutting the stone into, say, 100 pieces, they would still be of unusually large size, larger than a man, but at least could have been stacked into a wall much more easily. One is left with the unsettling thought that the reason they used these huge stones was because they could use them—and do it relatively easily, though today we have no idea how.

Ragette makes one final interesting comment on Baalbek:

> The real mystery of Baalbek is the total absence of written records on its construction. Which emperor would not have wanted to share the fame of its creation? Which architect would not have thought of proudly inscribing his name in one of the countless blocks of stone? Yet, nobody lays claim to the temples. It is as if Heliopolitan Jupiter alone takes all the credit.[39]

Osirian Remains in Egypt

If no one (except Nimrod in Arab legend) wants to claim Baalbek for themselves might it be a remnant of the Osirian civilization, one of those pieces that Isis sought after? Other vestiges of Osiris still exist in the eastern Mediterranean. The foundation ashlars of the Wailing Wall at Jerusalem are also gigantic blocks said to be similar to those at Baalbek. A massive block at the base of the Western Wall of the Temple Mount can now be seen and photographed inside an archeological tunnel excavated by Israeli archeologists. We do not know all the dimensions of this huge block of stone, but it is similar in size to those at Baalbek.

Megalithic ruins found under water at Alexandria, Egypt are also believed to predate the dynastic Egypt of the Pharaohs. The submerged megalithic ruins are another clue to ancient Osiris. Alexandria is not really an Egyptian city, it is Greek. As one might easily guess, Alexandria is named after Alexander the Great, the Macedonian king who first conquered the city-states of Greece in the 3rd century BC and then set out to conquer the rest of the world, starting with Persia. Persia was also Egypt's traditional enemy, and so Egypt fell willingly into Alexander's hands. He went to Memphis near modern-day Cairo and then descended the Nile to the small Egyptian town of Rhakotis. Here he ordered his architects to build a great port city, what was to be Alexandria.

Alexander then went to the temple of Ammon in the Siwa Oasis where he was hailed as the reincarnation of a god, which is to say, some great figure from ancient Osiris or Atlantis. Which god, we do not know. He hurried on to conquer the rest of Persia and then India. Eight years after leaving Alexandria, he returned to it in a coffin. He never saw the city, though his bones are said to rest there to this day (though no one has ever found the tomb).

Of all the mysteries of Alexandria, however, none is more intriguing than the megalithic ruins which lie to the west of the Pharos lighthouse near the promontory of Ras El Tin. Discovered at the turn of the last century by the French archaeologist M. Jondet and discussed in his paper "Les Ports submerges de l'ancienne Isle de Pharos," the prehistoric port is a large section of massive stones that today is completely submerged. Near it was the legendary Temple of Poseidon, a building now lost, but known to us in literature.[22]

The Theosophical Society, upon learning of the submerged harbor of megaliths, quickly ascribed it to Atlantis. Jondet theorized that it might be of Minoan origin, part of a port for Cretan ships. E.M. Forster theorizes, in his excellent *Alexandria: A History & a Guide*,[22] that it may be of ancient Egyptian origin, built by Ramses II circa 1300 BC. Most of it lies in 4 to 25 feet of water and stretches for 70 yards from east to west, curving slightly to the south.

Probably the true origin of the massive, submerged harbor, which was definitely at least partially above water at one time, is a blend of Jondet's theory of Minoan builders and the Theosophical Society's belief that it is from Atlantis.

In theory, with the Mediterranean slowly filling up with water, the sea would have stabilized after a few hundred years, and then the remnants of the Osirians, using a technology similar to that of Atlantis, built what structures and ports they could. Later, in another tectonic shift, the port area (probably used by pre-dynastic Egyptians) was submerged, and was then essentially useless.

It is interesting to note, with regard to this theory, that a temple to Poseidon was located at the tip of Ras El Tin. Atlantis was known to the ancients as Poseid, and "Poseidonis" or "Poseidon" was a legendary king of Atlantis. Similarly, Poseidonis and Osiris are thought to be the same person. The main temple at Rhakotis, the Egyptian town which Alexander found at the ancient harbor, was naturally dedicated to Osiris.

What we are learning about the megalithic masterminds is that their buildings occur all over the world, and many of them are underwater and difficult to reach!

The Amazing Megaliths of the Andes

At a leveling-off of a hill overlooking the Cuzco Valley in Peru, is a colossal fortress called Sacsayhuaman, one of the most imposing edifices ever constructed. Sacsayhuaman consists of three or four terraced walls going up the hill and the ruins include doorways, staircases and ramps.

Gigantic blocks of stone, some weighing more than 200 tons (400 thousand pounds) are fitted together perfectly. The stone blocks are cut, faced, and fitted so well that even today one cannot slip the blade of a knife, or even a piece of paper, between them. No mortar

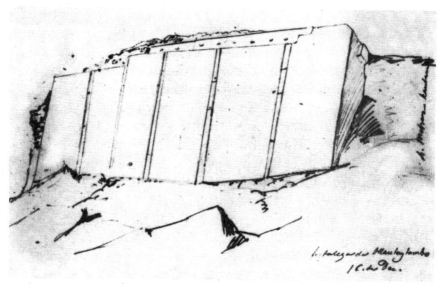

A drawing from 1843 of some of the megalithic blocks at Ollantaytambo.

is used, and no two blocks are alike. Yet they fit perfectly, and it has been said by some engineers that each individual stone had to have been planned well in advance; a twenty-ton stone, let alone one weighing 80 to 200 tons, cannot just be dropped casually into position with any hope of attaining that kind of accuracy! The stones are locked and dovetailed into position, making them earthquake-proof. Indeed, after many devastating earthquakes in the Andes over the last few hundred years, the blocks are still perfectly fitted, while the Spanish Cathedral in Cuzco has been leveled twice.

Even more incredibly, the blocks are not local stone, but by some reports come from quarries in Ecuador, almost 1,500 miles away! Others have located quarries a good deal closer, only five miles or so away. Though this fantastic fortress was supposedly built just a few hundred years ago by the Incas, they leave no record of having built it, nor does it figure in any of their legends. How is it that the Incas, who reportedly had no higher mathematics, no written language, no iron tools, and did not even use the wheel, are credited with having built this cyclopean complex of walls and buildings? Frankly, one must literally grope for an explanation, and it is not an easy one.

When the Spaniards first arrived in Cuzco and saw these structures, they thought that they had been built by the devil himself, because of their enormity. Indeed, nowhere else can you see such large blocks placed together so perfectly. I have traveled all over the world searching for ancient mysteries and lost cities, but nowhere else have I seen anything like this!

The builders of the stoneworks were not merely good stone masons—they were beyond compare! Similar stoneworks can be seen throughout the Cuzco Valley. These are usually made up of finely-cut, rectangular blocks of stone weighing up to perhaps a

Lord Kon Tiki Viracocha at Tiwanaku.

ton. A group of strong people could lift a block and put it in place; this is undoubtably how some of the smaller structures were put together. But in Sacsayhuaman, Cuzco, and other ancient Inca cities, one can see gigantic blocks cut with 10 or more angles on each one.

At the time of the Spanish conquest, Cuzco was at its peak, with perhaps 100,000 Inca subjects living in the ancient city. The fortress of Sacsayhuaman could hold the entire population within its walls in case of war or natural catastrophe. Some historians have stated that the fortress was built a few years before the Spanish invasion, and that the Incas take credit for the structure. But, the Incas could not recall exactly how or when it was built!

Only one early account survives of the hauling of the stones, found in Garcilaso de la Vega's *The Incas*.[40] In his commentaries, Garcilaso tells of one monstrous stone brought to Sacsayhuaman from beyond Ollantaytambo, a distance of about 45 miles:

> The Indians say that owing to the great labor of being brought on its way, the stone became weary and wept tears of blood because it could not attain to a place in the edifice. The historical reality is reported by the Amautas (philosophers and doctors) of the Incas who used to tell about it. They say that more than twenty-thousand Indians brought the stone to the site, dragging it with huge ropes. The route over which

55

Engineer Christopher Dunn examining stones at Ollantaytambo with precision tools.

they brought the stone was very rough. There were many high hills to ascend and descend. About half the Indians pulled the stone, by means of ropes placed in front. The other half held the stone from the rear due to fears that the stone might break loose and roll down the mountains into a ravine from which it could not be removed.

On one of these hills, due to lack of caution and co-ordination of effort, the massive weight of the stone overcame some who sustained it from below. The stone rolled right down the hillside, killing three- or four-thousand Indians who had been guiding it. Despite this misfortune, they succeeded in raising it up again. It was placed on the plain where it now rests.[40]

Even though Garcilaso describes the hauling of one stone, many doubt the truth of this story. This stone was not part of the Sacsayhuaman fortress, and is smaller than most used there, according to some researchers, although the stone has never been positively identified. Even if the story is true, the Incas may have been trying to duplicate what they supposed was the construction

56

technique used by the ancient builders. While there is no denying that the Incas were master craftsmen, if one credits this tale one would have to wonder how they would have transported and placed the 100-ton blocks so perfectly, given the trouble they had with only one stone. It is curious that Garcilaso would say that the stone came from Ollantaytambo at the north end of the Sacred Valley. It seems unlikely a stone would be brought from such a distance to Cuzco, however Ollantaytambo is the location of many of the "lazy stones" that have been left along ancient roads and ramps. These huge rectangular blocks of red granite known as Andesite— weighing from 60 to 100 tons—were meant to be fitted into walls at the so-called Sun Temple on the edge of a cliff on the northern edge of the small town of Ollantaytambo. Some simply did not make it all the way from the quarry across a river and up an artificial ramp to the spectacular building site. These are called the lazy stones and why work suddenly stopped or these stones only made part of the journey has never been explained.

One of the problems with moving such large stones is that only a few hundred people could be pulling on ropes while they wore harnesses to drag a huge block of stone, and it is not known where these people would stand when blocks have to turn a corner on an artificial ramp.

Jean-Pierre Protzen and the Mystery of the Stone

The mystery of the building of Ollantaytambo is one that has a profound impact on modern archeology. Not only is it a mind-boggling and impressive site, but allegedly it was built by a culture that was lacking much of the knowledge that cultures in Europe, Africa and Asia in the Old World had. Mainstream archeologists, if their theories which were now expounded in books and universities all over the world were correct, would have to be able to show that the Incas built Ollantaytambo by simple means only a few hundred years before the Spanish arrival. So, how was it done?

The answer, according to these mainstream voices, is by simple brute force. In the case of the cutting and articulating of the blocks in such a "perfect" and difficult fashion, it was by tedious, time-consuming beating with primitive stone hammers and soft metal chisels until the perfection was achieved. It was with "patience,

patience, patience" and more and more brute force that a gigantic block was dragged and then lifted into place. The polishing and final cutting of the blocks was even more time-consuming and tedious detailed work on very hard stone. And so the Incas (or whoever) made these structures. Never mind the reason they would choose to do something so incredibly difficult by our standards.

Yes, even the conservative archeologists are impressed by the building of Ollantaytambo, and so they must offer some reasonable explanation of how all this was done. The quarry for the stones is on a mountainside across the river and as we have noted, a trail of "lazy stones" can actually be followed down to the river and then to the western side of the sharp mountain ridge upon which the Sun Temple sits. The remains of a ramp up the west side of the ridge and other roads can be also be seen. Still, the stones had to be moved down a mountain, across a river and then up the steep slope of a ridge to a small plaza with steep drops on every side. Every tourist must ask himself how it was done... and the usual answer is not entirely satisfying.

One person to turn to for an expert's opinion on the matter is Jean-Pierre Protzen, a Swiss architect/stonemason who teaches at the University of California at Berkeley. Because of his detailed investigation into megalith building in the Andes, he is the best source for any mainstream explanation for the building of Ollantaytambo, Sacsayhuaman, or other super-megalithic sites. Protzen generally tackles the problems from a practical point of view, and he says he goes along with mainstream archeologists who claim that the Inca are the builders of Cuzco, Sacsayhuaman, Pisac, Ollantaytambo and Machu Picchu—but that an earlier Tiwanaku culture built the massive ruins of Tiwanaku and Puma Punku, as well as perhaps the towers at Cutimbo and Sillustani. However, he admits that some things remain a mystery to him and he is baffled by the keystone cuts found at Ollantaytambo and the Qoricancha, which he knows were used at Tiwanaku and Puma Punku.

Protzen discussed his early findings in a report in *Scientific American* (February, 1986). In 1993 his findings were published in the book *Inca Architecture and Construction at Ollantaytambo,*[50] published by Oxford University Press. This book was also published in Spanish as *Arquitectura Y Construccion Incas en Ollantaytambo* in Peru in 2005, and can be found in some bookshops in Lima or Cuzco.

Protzen is currently a professor at the University of California at Berkeley but originally got a diploma in architecture at the University of Lausanne in Switzerland. Said to be an extraordinary researcher, he is interested in "design theory and methods, Inca architecture, and construction techniques." He has received honors that include research fellowships from the Swiss National Science Foundation and the University of California, and an International Architecture Book Award. He currently teaches courses on design theories and methods, logics of design, and research methods.

Fortunately for those of us interested in ancient technology in Peru and Bolivia, he is also interested in the subjects of "the logics of design, design, planning, and construction principles of ancient civilizations, particularly Pre-Columbian South America." His books and research are very important. Let us look at what he has to say.

In his 1993 book Protzen addresses, as best he can, all the major issues at Ollantaytambo, including the quarrying, transporting and cutting and dressing of the stones. He also concludes with a short but interesting chapter on the "chronology" of Ollantaytambo. Why should there be a chronology of this site if it was built by the Incas? Well, because the keystone cuts and the clamps or "cramps" as he calls them (the more technical word) are an indication, which he does not deny, that the builders of Tiwanaku and Puma Punku were also the original builders of Ollantaytambo. This of course upsets the traditional archeological dogma that all the monumental sites in the Sacred Valley and around Cuzco were built by the Incas many hundreds if not thousands of years after the building of Tiwanaku.

Since Protzen is trying to defend the status quo his chronology chapter is a cautious check into the discovery of stone blocks with keystone cuts in them, both at Ollantaytambo and at the Qorichancha-Temple of the Sun in Cuzco, another building which may be pre-Inca. This is a sensitive topic for Protzen, as he wants to be seen as part of the establishment and mainstream, since his university and academic career may be at stake. Protzen says:

An argument persists that the Wall of the Six Monoliths

59

[the "Sun Temple" at Ollantaytambo] and the vanished structures from which blocks have been recycled predate the Incas and were the works of earlier Tiahuanaco culture. Support for the argument is found in the step motif carved on the fourth monolith and in the T-shaped sockets cut into several blocks, both believed to be hallmarks of Tiahuanaco-style architecture. Even Ubbelohde-Doering could not help but be reminded of Tiahuanaco when looking at these details and the bond in the First and Second Walls. He did, however, explicitly write that there were no reasons to believe that any of these structures at Ollantaytambo predated the Incas.

A variant of this argument is that Tiahuanacoid elements were brought to Ollantaytambo by *Qolla mitmaq* stonemasons—that is stonemasons from around Lake Titicaca ...the presence of *Qolla mitmaq* at Ollantaytambo is historically documented. The only question here is why stonemasons from around Lake Titicaca should have remembered anything Tiahuanacoid when for several centuries nothing like it had been built.

If anything at Ollantaytambo reminds me of Tiahuanaco it is neither the step motif nor the masonry of the First and Second Walls, but the T-shaped sockets and regularly coursed masonry of strongly altered andesite. The step motif was widely used by the Incas—at Ollantaytambo it appears also on the splash stone of the Bano de la Nusta fountain—and there is no reason to believe it derived from Tiahuanaco. The bond of the first and second walls is more Inca-like than it is Tiahuanacoid, as will be shown.

Many T-shaped sockets are indeed found at Tiahuanaco, in particular at the site of Pumapunku, where some still line up back to back on adjacent blocks. From Tiahuanaco it is known with certainty, since many cramps have been retrieved, that copper cramps inserted into the sockets held the building blocks together... cramp sockets were found on loose building blocks retrieved from the church and monastery of Santa Domingo at Cuzco after the earthquake of 1950. In Cuzco, as at Tiahuanaco, the shape of the cramps was not limited to T; they were also made in the U, Lorrain cross, double T, and other shapes. The cramp sockets at Ollantaytambo, however, are exclusively T-shaped. Neither in Cuzco nor at Ollantaytambo were the sockets found on blocks in situ, all blocks with sockets have been moved out of their original context, and, as mentioned before, no cramps have ever been found. Unfortunately, these facts confuse rather than shed light on the questions of who the builders were who used cramps in their construction in Cuzco and Ollantaytambo and of where the original buildings stood.

The masonry of green, strongly altered andesite, with its blocks of perfectly flat wall faces and regular coursing of ashlars of equal height, bears a striking resemblance to masonry at the Pyramid of Akapana at Tiahuanaco, for example. Even the best of regularly coursed masonry in Cuzco shows minor variations in height within a course, resulting in wavy horizontal joints, and even the smoothest of walls reveals traces of sunken joints. The same holds for the First and Second Walls of Ollantaytambo. Sunken joints are a direct consequence of the dressing technique used by the Incas, and the wavy courses are due to the one-on-one fitting and laying technique. What, then, is intriguing about the masonry of strongly altered andesite, as exemplified in Llanos' wall of twelve ashlars, is that it suggests different dressing, fitting, and laying strategies. Stones may have been prefitted on the ground and subsequently hoisted into place on the wall. Evidently, such a technique is most profitably used in regularly coursed masonry with all ashlars exactly the same height. A careful study of the orderly rows of blocks excavated by Llanos may indeed reveal the secret of how the strongly altered andesite blocks were cut and assembled. Unfortunately, too much debris has again accumulated between and over these blocks, which cannot be investigated in detail without reexcavation.

If we assume that a different construction technique was used to erect the walls of green stones, questions arise as to whether this technique is a late development, a refinement of an older technique, or an old one, possibly predating the Incas. And was the masonry of green stones contemporaneous with rhyolite masonry in which T-shaped clamps were used? With the evidence at hand, these questions cannot be answered.[51]

So, in academic speak, we have Protzen approaching the sensitive topic of whether the presence of the keystone cuts at Ollantaytambo and the Qoricancha in Cuzco are evidence that many of the structures in Peru that are attributed to the Incas are probably actually pre-Inca in origin. He is basically saying "yes" to this question but with the caveat that "with the evidence at hand, these questions cannot be answered." Here he uses the term "clamp" instead of "cramp." In the Spanish version of his book clamps and cramps are called "grapas."

Protzen starts to ask the right questions, but then suddenly stops, because the obvious conclusions would lead him far from the current academic dogma—and he knows he cannot go there! The crux (Andean crux, if you will pardon my pun) of the matter is this, and it is clearly stated by Protzen: keystone cuts and the clamps that go with them (although no clamps were ever discovered in Peru, he says) are associated with Tiwanku and Puma Punku. So, he admits, it is not some "wacky" idea that stones with keystone cuts in them—found at Ollantaytambo and Cuzco—should be associated with Tiwanaku.

But there is a big problem here, which Protzen admits. Tiwanaku culture flourished about one thousand years before the Inca Empire and its supposed building of the structures attributed to it today. So... how to explain this?

Protzen, to his credit, brings up the standard explanation in archeological circles: the builders of Ollantaytambo were brought from the Lake Titicaca area where they must have seen the use of keystone cuts at Tiwanaku and Puma Punku and then brought them to Cuzco and Ollantaytambo. But Protzen isn't buying this argument at all. He says, "The only question here is why stonemasons from around Lake Titicaca should have remembered anything Tiahuanacoid when for several centuries nothing like it had been built."[51]

This is completely the point, and Protzen totally gets it, while other mainstream archeologists don't seem to quite comprehend

A photo from Protzen of some of the keystone cuts on blocks in Cuzco.

this important fact: keystone cuts are an undeniable fact at the megalithic sites of Cuzco and Ollantaytambo, but they were not used by the Inca. What are they doing there? They must have been made before the Incas arrived, just like all the other buildings in the vicinity of Cuzco.

Even given the evidence of the keystone cuts, the scholars were unwilling to change their story. Whether or not keystone cuts were associated with Tiwanaku, built by a different culture hundreds of years before, the structures around the Sacred Valley had to have been made by the Incas. It does not make sense, but it is the academic dogma and Protzen has no real choice but to somehow stick with it, otherwise he could lose his job and academic standing. Therefore he says that with the evidence at hand he just cannot figure it out.

Protzen points out that no copper or bronze clamps or "cramps" have been found either at Ollantaytambo or Cuzco. Does he think that some stonemason cut these careful sockets into the hard granite for no reason? Was no molten metal clamp ever to have been poured into the keystone cut "Tiwanaku-style"? The obvious answer, which Protzen probably realizes, is that the missing clamps were taken by people who happened upon the ruins over the years, and melted down for other purposes. This would explain their absence, particularly in Peru where the Spanish, who were known to be very

63

A granite block with saw marks at the quarry above the town of Ollantaytambo.

fond of metals, were doing a whole lot of looting.

It is interesting to note here that no dating technique was undertaken by Protzen to try to solve the mystery of the keystone cuts. This is because there is no way of dating cut stone blocks at present. Any dating must be in the "context" of the site itself and objects other than stones must be dated. We have seen the pitfalls of this dilemma, and it is to Protzen's credit that he avoided that course.

Protzen strangely ends his book with this brief and mysterious admission. Keystone cuts and everything associated with them are from the Tiwanaku culture, but the Incas built these structures hundreds of year later—so I cannot figure it out, he says. I need more data. *Voila*, Protzen keeps his academic job, but still manages to salt the waters with his own vague opinion that the Incas must have inherited some of these structures and used them for their own purposes. But this is not all that we can glean from Protzen's interesting book. His careful examination of the Ollantaytambo quarry and his theories on how the stones were moved are just as important.

The Moving of the Stones from the Quarry

The *Kachiqhata* quarry is high on the hill above the Urubamba River and is not a normal quarry in the strict sense, where stones are removed from bedrock by undercutting them, or cut out from a cliff or other rock face. It is an area of huge rockslides where giant boulders of granite have fallen from cliffs above. There are different sized rocks, from small to gargantuan. It was here that the builders of Ollantaytambo—whoever they were—came to carefully select their largest stones.

There are three main slides of rocks and a road begins at the bottom of the central slide. It is apparent that from here the squared rectangular blocks of red granite were moved down the mountain, first on a road that went slightly downhill while traversing the hillside to the south. This is the same road (or trail) we had just taken to the quarry, starting at the bridge in town; but the giant blocks would not have been taken all the way to the bridge. There is a spot in the road, approximately opposite the site of the Sun Temple, where the blocks would have been pushed off, down a big slide, to land on the west side of the river.

Protzen says that the largest of the stones from the quarry weighs an astonishing 106,000 kiloponds. A kilopond is also known as a kilogram-force and is a gravitational metric unit of force. It is equal to the magnitude of the force exerted by one kilogram of mass in a 9.80665m/s^2 (meters per second squared) gravitational field, which is what is known as "standard" gravity. One kilopond or kilogram-force is approximately 2.204622 pounds. A stone that was 106,000 kiloponds would weigh an astonishing 233,200 pounds or 116 tons. This is the weight of the stone that Protzen would be using for his calculations on moving the stones in his book (some of which we will see shortly). It is of considerable size and weight, and Protzen is able to make some impressive calculations—but again he cannot completely figure it out! Let us look at his important study on moving the stones down the mountain from the quarry, across the swift-flowing Urubamba River, and then to the mountain ridge where the remains of a ramp can be seen.

Says Protzen:

> Not counting those remaining in the quarries near *'Inkaraqay,* some forty blocks of coarse-grained rose rhyolite [lying between the quarry and the Sun Temple] signal the way over which the stones were transported from the quarries to the construction site of the temple area at the Fortress of Ollantaytambo. All blocks show signs of having been worked. From the southern quarry, the blocks traveled eastward over ramps to slides leading down a ravine and over a road to *'Inkaraqay,* where they joined the blocks coming from the western and northern quarries. From here, all blocks were hauled over a road to the last slide, plunging toward the Urubamba River. One block is still stuck in the slide, and four more are dispersed on the alluvial plain between the bottom of the slide and the river. Patches of differential growth patterns in the crops raised on the alluvial plain may indicate that more blocks, no longer visible, are buried there.
>
> There are indications that the last slide was not always in use or that it became a shortcut not originally intended. At the head of the slide, an old ramp, the continuation of the road above, turns east; it can still be traced for some 250 meters before it runs into a landslide that has wiped all visible remains. If we project the course of the ramp from where it ends today, it may have reached the river just west of *Runku Raqay.* Here, the river emerges from a canyon. The river is still narrow and deep, the left bank is very steep, and several terraces on the right bank bar the way, making this an unlikely ford for blocks weighing tens of metric tons. It is thus possible that the ramp was abandoned because it was impractical or simply because the slide was more expedient.
>
> The last slide points exactly at a large abandoned block on the opposite riverbank, just beyond the railroad tracks, suggesting that the river was forded by the direct prolongation of the slide. Harth-terré argued that the blocks were moved across the river near an island in the river that

is about 600 meters downstream from the slide. Considering the tremendous forces of the rushing Urumbamba River, swelled by the heavy summer rains, one may question whether that island even existed 500 years ago. As Harth-terré noted, the Incas had canalized the Urubamba River along this particular stretch, leaving the river to find its own course. My view is that the location of the fording must have changed over time, perhaps as a function of the river's flow and the conditions of its bed and banks after the summer rains. I am led to this view because the four blocks on the left bank and the twenty or so blocks on the right bank are relatively widely scattered, suggesting that no preestablished path was followed. Also, I have not been able to detect the slightest indication of a roadbed linking the riverbank with the foot of the ramp leading up to the Fortress. On the left bank, a causeway is intimated by a differential growth pattern in the crops. Since it does not line up with the slide, it is not clear what purpose this causeway, if it existed, may have served.

The volume of water carried by the Urubamba River is from three to four times larger in the wet season than it is in the dry season, when the water level in the area of the fording reaches about 1 to 1.2 meters. It seems reasonable to assume, then, that the hauling of blocks across the river was not a year-round operation but was limited to the winter months, when water levels were low.

As mentioned earlier, once across the river, the blocks were moved along an undetermined path to the base of the Temple Hill and up a long inclined plane with a slope of 8° to the temple area. The last stretch of this plane is supported by a formidable retaining wall 16 meters high.[51]

So here we have some fascinating information from Protzen who has done a pretty good job of finding the route that the stones took, although he does not address how they got across the river. Certainly the builders would want to do it when the river was at its lowest point, but he still does not know how. It seems they did not build a bridge. No bridge made of wood could probably hold the largest blocks, anyway. They must have been either dragged or "flown" across the river.

It is curious that so many blocks made it across the river, but were abandoned on the eastern side of the river. One would think that the hard part had been done, getting them across the river, yet Protzen says that 20 were abandoned before reaching the fortress. All these abandoned blocks of red granite (rose rhyolite) are known as the "lazy stones" and there clearly are quite a number of them—40 in all, according to Protzen, not counting some that are now buried.

These "lazy stones" are all over the place. At the very start of the four slides of the quarry, a number of gigantic blocks fully dressed and squared lay ready to make the trip. A number lie along the upper road to the slides, and some made it down to the western riverbank. A few giant blocks lay on the eastern side of the river going up the bank, some more lie in the fields on the way to the fortress and several blocks are beginning to make the ascent up the ramp to the small plaza. Even there, among blocks that have been actually put into place a few lazy stones can be found.

It is mystifying why so many blocks of stone were apparently abandoned along the way to the temple site when so much effort had been expended in preparing them and moving them as far as they got. But what really boggles the mind here is that any of them actually made it to the narrow ridge of the Sun Temple at all! How did they maneuver giant blocks across the ground, up to a small plaza on a sharp, cliff-faced ridge on the edge of the ancient town of Ollantaytambo? Protzen is determined to figure it out and the first thing that he notices are "drag marks" on most of the blocks. Says Protzen:

> How were the blocks, some of which weigh in excess of 100 metric tons, transported over 5 kilometers from the quarries, across the river and up to the Fortress? I found the first clue to this problem on block 29 on the southwest side of the Sun Temple, on which one observes a smooth, yet uneven, polish traversed by fine, more or less parallel striations. This polish, I contend, results from dragging the blocks over the bare surface of the roadways. Inspecting the polished face of this block, one notices that the polish extends

over only the prominent portions, not the depressions, of the face. Close inspection of the recessed surfaces reveals sharp boundaries between the polished and the nonpolished surfaces on the opposite end. From this particular feature, it is possible to infer the direction in which the block was dragged. The sharp edge of the recessed part was the leading edge; the blurred part was the trailing edge. As the block moved forward, the surface material on the road ground against the protruding portion of the block; the material escaped into any depression, only to accumulate at the back of the depression, where it got compressed and ground between the roadbed and the block, so that the depression's trailing edge was worn down.

Some of the abandoned blocks along the road from the quarries to the Fortress were buried too deep to have all their faces inspected, but all other blocks have at least one face with polish and striations. Drag marks are still detectable on many wrought stones strewn about the temple area. As one would expect, drag marks are conspicuously absent on blocks still in the quarries. Drag marks are always found on the broadest face of a block, indicating that blocks were transported in their most stable position. On some blocks, drag marks can be found on two opposite faces, suggesting that the blocks were turned over during their transport, perhaps when tumbling down a slide or when fording the river.

The presence of drag marks does not, of course, exclude the possibility that the block were moved some other way, at least along parts of the road. It is imaginable that the drag marks resulted only from the blocks slithering down the slides, although I doubt that the slides were long enough to produce such extensive marks as those observed on most blocks. Not only were the slides hollowed out in a troughlike manner, but so were the ramps in the quarries. The hollowing can be explained by the successive passages of blocks, each one carving out a little bit more of the road surface.[51]

What is interesting to note here is that Protzen is saying that drag marks can clearly be seen on the stones that were moved, and those drag marks could be the result of these cut and dressed stones going down the several slides that descend from the quarries to the Urubamba River. He does not think the slides alone would have produced the marks he sees, but admits it is possible—this could be important as we will see later. But how did these super-heavy stone blocks get moved along the trails to these slides, and then across the river and up the other side? Protzen thinks they were dragged, which would have accentuated the drag marks to a degree consistent with what he observed. This must have been a tremendous engineering feat and required what would seem to be superhuman effort.

Indeed, a calculation can be made to ascertain the number of humans pulling on ropes that would be necessary to attain the force needed to move the stones. And Protzen did this—the answer that he came up with was that 1,800 people pulling on an elaborate rope harness could drag a gigantic block that was held in a rope webbing that gripped the stone.

How was all this done—and where did the 1,800 people stand in certain difficult places, such as the final complicated phase of getting the blocks onto the small plaza on the ridge? Protzen addresses that problem later, but first let's review the basic calculations:

> If the blocks were, in fact, dragged with ropes over the bare surface of the roads, this raises such questions as: How many people did it take to drag a block of 100 or more metric tons? How were the people harnessed to the blocks? How were the ropes attached to the block? How were the blocks maneuvered around curves and corners?
>
> Answers to the last three questions depend on uncovering descriptions of the transportation technique far more detailed than the ones [discussed] above or on finding material traces on the blocks themselves. The first question is independent of such evidence, since it lends itself to analytic treatment. The force required to drag any block is given by the equation

$$K = f \bullet P \bullet cos \ a \pm P \bullet sin \ a$$

where K is the required force, f is the coefficient of friction, P is the weight of the block, and a is the slope of the ramp. The + is used to compute the forces to pull uphill; the −, to drag downhill. The weights of the blocks and the slopes of the ramps are unknown. The coefficient of friction depends on the surfaces of both the blocks and the ramps. What the surface of the ramps was in Inca times is not known. The current conditions suggest two possibilities; the ramps either were surfaced with a ballast of broken rocks or were finished with compacted dirt. Since none of the standard physics handbooks offer adequate coefficients for either of these conditions, it was necessary to establish empiracally appropriate values for f. I built two surfaces approximating the two road surfaces encountered today at Ollantaytambo. With the aid of a dynomometer, I dragged a block of 42 kilograms repeatedly over these surfaces. The average results were f = 0.75 for compacted dirt, and f = 0.7 for a ballast of broken rock. The latter was somewhat smaller than the former because the loose broken rocks had, to some extent, the effect of ball bearings.

Assuming the roads were made of a ballast of broken rock, computations yield the following values for the requisite forces to drag a block along the different stretches of ramps from the quarries to the construction site at the fortress. The slope at which friction is overcome by the sheer weight of the block—that is, at which the block starts to slide downhill on its own—is approximately 38°. This is very close to the slopes of the various slides along the transportation route. A small impulse was all it would take to send a block skidding down a slide. Down a slope of 8°— the approximate incline of the road near '*Incaraqay*—the necessary force to move a block would be equal to about 55 percent of its weight; along a flat stretch it would be 70 percent, and up a slope of 8°, the approximate incline of the ramp to the fortress, it would be 84 percent. Thus to drag uphill the largest block (which weighs 106,000 kiloponds) remaining in the quarries of Kalchipqhata, a force of 89,000 kiloponds would be required. Assuming that a person can pull consistently with a force of 50 kiloponds, it would take some 1,780 people to accomplish the job. This may seem like a very large number of people, but I find a certain consistency between this number and Cieza de Leon's count:

> Four thousand of them were breaking
> stones and extracting stones;
> six thousand were hauling them
> with big ropes of hide and leaf fibers...

Although in the passage Cieza was writing about the construction of the fortress of Saqsayhuaman, not that of Ollantaytambo, I see no reason why one could not assume that the Incas used similar building practices at all their highland sites. The moving of heavy stones by 1,800 workers at Ollantaytambo is commensurate with the long-distance transportation of heavy and bulky objects in other ancient cultures in both the Old and New Worlds. One need only be reminded of the orthostats of Stonehenge, the colossal statues of Ramesses II at Luxor, the enormous Idol of Coatlinchan at Teotihuacan, or the stone giants of Easter Island to appreciate the remarkable achievements of ancient transportation engineers. A tomb painting at El Bersheh, Egypt, and a limestone tablet from Nineveh depict the dragging of heavy statues by scores of men tugging on long ropes hitched to sleds under the statues. It is not clear whether these representations of ancient transports represent the actual number of draggers or whether only "an impression of a great crowd of men drawing on the ropes was intended." Engelbach estimated that a work force of 6,000 men would have been required to move the unfinished obelisk at Aswan.

To reduce the work force required in hauling, one has to invent a mode of reducing friction or of increasing the effectiveness of the work force's output. Short of the wheel, which the Incas did not know, there are several other possible ways of reducing friction, including the use of either rollers or skids and lubrication.

The use of rollers is frequently mentioned in the literature, but the evidence in support of this hypothesis is meager and certainly not conclusive. Skids can be used in two ways: they can be fixed to the object to be moved, or they can be placed on the roadbed like railroad tracks. Outwater argued that the latter method was used in conjunction with rollers at Ollantaytambo. Unfortunately, the evidence for the use of skids does not fare any better than that for the use of rollers; at best, it is incidental. In an experiment conducted in 1986, a group of eight workers at Ollantaytambo were asked to move a 1.5-ton building block without instructions on how to do it. They resolved the problem by laying down two long poles, tracklike, over which they pulled the block with ropes while pushing it with levers from behind.

The use of lubricants is illustrated in the El Bersheh painting. A man standing at the prow of a sled is pouring a liquid into the path of the sled. The evidence for the use of a lubricant, probably wet clay, at Ollantaytambo is merely circumstantial. A stone-polishing experiment, to be

An artist's conception of the building of the megalithic walls at Ollantaytambo.

described in detail in Chapter 11, revealed that to produce a highly polished surface, an exceedingly fine abrasive was needed. It is thus unlikely that the polish observed on the dragged blocks was brought about by the coarse road ballast, unless it was mixed in with substantial amounts of very fine sand or clay. If clay was used on the roadbeds, wetting it would have helped the blocks to move along. It would have reduced the coefficient of friction from 0.7 to about 0.2, a change that could have reduced the requisite transportation crew to 715 people.[50]

So, we can gather from Protzen's detailed analysis that it might have only taken 715 people to haul the largest blocks at Ollantaytambo if some kind of sled (and a system of wetting the soil, as the Egyptians were known to do) had been used in the dragging of the blocks. However, he does not think that they used such a technique, nor any kind of roller or lever.

The drag marks observed by Protzen seem to be on all the stones that were removed from the quarry. During the slides that each block would have to have made down to the river it would have acquired a certain type of drag mark that would also be similar to the drag mark made on the stone if it were being dragged across a level road or slight slope. If the megalithic block were being forced forward by levers, then a different type of drag mark would also be seen on the block. Protzen could not find this type of mark. Nor could he see how a large number of people deployed along a mountain road were able to negotiate turns without the pulling team overshooting the turn and dragging the block all the way to the point of the turn before turning back and rehitching in a new direction. The problem at Ollantaytambo is that at several of the critical turns, there is no room for the team to keep going straight—they would fall off the mountain.

Says Protzen:

In the transport scenes on the Assyrian tablets, the sledges are not only pulled on long ropes, but also shoved from behind with long levers. The Assyrians were obviously aware of the amplifying effect of levers, even if their way of

The Nineveh tablet showing a large statue being pulled by ropes.

pulling the levers down (shown on the tablets) was not the most effective and if at the angle at which the levers were applied the load was more likely to be heaved rather than pushed forward. To get the most forward push, the levers had to be kept as vertical as possible, and to get the most leverage, the pull had to be exerted at the very top of the lever. The short sides of the largest blocks at Ollantaytambo are wide enough to receive at least two levers. Assuming a leverage ratio of six to one and about sixty workers to pull on each lever, the workers cold produce a combined effective initial shoving force of some 35,880 kiloponds. The output of the lever crews is thus about 40 percent of the calculated 89,040 kiloponds required to move the block discussed earlier uphill on a slope of approximately 8°. Accordingly, the pulling crew could be reduced by 40%, from 1,800 to 1,080 people. Adding to that the lever crew of 120 people, one is left with a total transportation crew of 1,200.

One could imagine that to harness the pulling crew to a block, a big net, into which four or more long ropes were woven, was thrown over the block's top and sides. As shown on the Nineveh tablet, each rope would have had a series of twin loops attached to it at 80-centimeter intervals for two men to pull, one on each side of the rope. To keep the ropes at regular intervals, a yoke of some sort would have had to be mounted at the front of the block. Assuming that the four long ropes were spaced about 160 centimeters apart. 1,800 people could have been deployed along the ropes, 450 to each, 225 on either side. In this fashion, the whole train would have stretched over 187 meters [the length of two

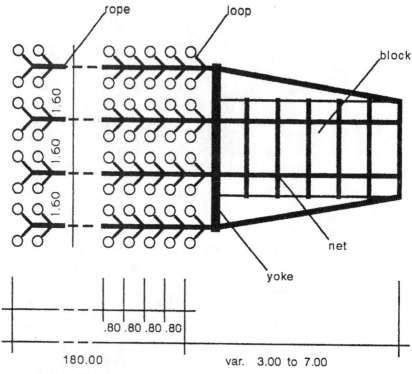

The harness proposed by Protzen to move the stones at Ollantaytambo.

football fields] along the roadway.

If I have any reservation about my explanation of the technique of dragging at Ollantaytambo, it is my inability to propose a plausible way of how a large number of people deployed along a mountain road (at most 6 to 8 meters wide) negotiated turns, wide or sharp, without the pulling train overshooting the turns. Overshooting a turn would let the crew pull the block right up to the turn. Once there, the block could be pointed into the new direction with levers, and the crew rehitched to the front of the block. Unfortunately, the accidented [uneven] topography along the transport route at Ollantaytambo would not allow the pulling train to go beyond any of the critical turns.

The problem would not be resolved with smaller pulling crews. The reduced crew using wet clay or the crew aided by levers would have encountered the very same difficulties in negotiating turns. It appears that only crews working exclusively at the back of a block could properly approach and complete the turns. A combination of levers and lubrication could resolve the issue. The force put out by the crew working the levers would be sufficient to push the largest block uphill to the Fortress *if* the coefficient of friction was kept down to 0.2 or less. What I called *drag* marks would then be *shove* marks. The latter would be indistinguishable from the former, with perhaps one exception. The tendency of levers is to lift the blocks at the back, pushing the front into the ground. Thus the abrasion marks should be most pronounced at the front of the block and should gradually diminish toward the back. To date, I have not detected a fading of the abrasion marks on any block.[51]

So, Protzen is unable to find any "fading of the abrasion marks on any block" and therefore cannot find evidence that levers were used in the movement of the blocks. This lack of the use of levers seems to betray whatever ingenuity that the transport engineers might have had. In other words, it would seem that the planners and executers of this elaborate scheme to get gigantic blocks from a mountain rockfall across a river and up a mountain ridge, were not able to think of using things like rollers or levers. Yet, they were able to accomplish things that seem superhuman by even our modern standards! Was it all done with just the brute strength and sheer numbers Protzen is dutifully calculating for us?

This Swiss-American engineer is obviously fascinated by the how and why and when of the building on this amazing structure, as am I. Indeed,

A statue being pulled by ropes at El Bersheh, Egypt.

there are many such amazing structures in Peru and Bolivia and they contain many mysteries, but Ollantaytambo stands out, which is why Protzen has intentionally made a careful study.

How could this crew of 1,800 people have negotiated the narrow turns required to bring these gigantic blocks to the small plaza where the Fortress stands? Where did these people stand when they were dragging this massive block up to the sharp ridge of the Ollantaytambo fortress? Protzen cannot answer these questions. He has figured out how many men it would—at minimum—take to drag some 100-ton block of red granite, squared into a rectangle, up the hill to the Fortress plaza. But, he cannot figure out where all these people stood while doing so!

Did they build some gigantic ramp to allow this teeming team of haulers to pull with all their might on the intricate web of ropes and yokes that made the hauling of this massive block possible? Was this ramp later dismantled, and then the terraces now seen on the steep hillside constructed to grow crops? This seems highly unlikely and Protzen does not even mention the possibility.

Yet, this is exactly what various Egyptologists have proposed over the centuries to solve the mystery of the building of the Giza Pyramids, especially the Great Pyramid. Vast ramps going out from the pyramid, or in a spiral around the pyramid, have been conceived. These huge ramp complexes are then dismantled to the point that there is no trace of them left. This is not the case at Ollantaytambo, where the remains of a ramp going up the west side of the steep ridge can be seen. But this ramp could not accommodate the team described by Protzen, or the way it would have to have executed the curves.

In short, Protzen has done the basic calculations necessary to show what would have been needed to move these 100-ton blocks with brute force. One would need 1,800 laborers and a strong rope harness and network, including yokes, to drag these blocks over level or slightly sloped surfaces. But the solutions to such problems as getting the stones around corners, across a river or up onto a small plaza on a narrow ridge elude him. Such beguiling questions certainly add to the mystery of the construction of these impressive sites. How did these builders—Protzen presumes they were the Incas, though he seems to have his doubts—do this clever trick?

78

We assume that they were "primitive" in the sense that they did not know about the wheel or the use of levers, but they still managed to do something that we cannot yet comprehend. At least in this way, they are more clever than we are!

The Lazy Stones of Ollantaytambo

Since 1985 I have visited Ollantaytambo many times and I have never tired of visiting the site. I have only made the difficult journey to the quarries on the mountainside across the river from the town three times. But each time I have marveled at the many gigantic stones that have been dressed into huge granite rectangles that were to be moved down the mountain along specially made roads and slides and then across a river and up to the edge of a cliff. Along the way, on the edge of the road at various spots are the rectangular granite blocks that only made it part of the way to the so-called Sun Temple perched on the edge of a cliff. Why did they not make it? When did this work stop? Why were they working with such large blocks of granite and how would they get them across a raging river? No one has the answer to these questions.

Some of these "lazy stones" are quite large, while others are about half the size of the larger blocks that were successfully moved across the river and up to the small plaza at the fortress. One of the largest "lazy stones" was successfully dressed, moved down the slides, then across the river and right up to the final ramp on the west side of the fortress. But for some unknown reason it was never moved the final last bit uphill to the plaza. Why was all this effort expended on this stone—and the other "lazy stones"—when they were ultimately to be left on the side of the road? What is it that made these stones "lazy"?

Indeed, even the term "lazy" implies that there is something wrong with the stone, rather than with the transport crew that was "dragging" the stone. One theory on the moving of megalithic stones (to be discussed in greater detail in the following paragraphs) holds that the stones were not dragged at all, but moved by a type of levitation device that made the stones "leap" through the air. At Ollantaytambo, this process would involve applying the device to a stone to make it "jump" down the road from the quarry to the slides, where the stones would be pushed over the edge and retrieved at the bottom. They would again be made to jump to the river and across, and then up the slope to the fortress plaza. This process would produce the drag marks observed by Protzen, since the stones would go down the slides, and there would probably be some skidding action as the jumps were made. There would be no lever marks made, which is consistent with Protzen's observations. Perhaps during this process it was found that certain stones were "lazy" and could not be made to jump properly, and they were therefore abandoned where the process failed.

This fascinating "levitation" theory, described to me in a telephone conversation in the fall of 2003, claims that quartz crystals, when connected in a series and shocked with high voltages, will "bend." The source claimed that when a crystal is struck, put under pressure, or "bent," it will give off a piezoelectric signal and, incredibly, it actually loses the gravitational force that would naturally pull it toward the center of the mass (in this case, the Earth). The crystal then becomes essentially weightless, no matter how heavy it was before being bent by high voltages. If such an effect could be confirmed, then gigantic blocks of granite, which are full of small quartz crystals, could theoretically be moved with

One of the massive granite blocks at Ollantayambo, this one with a keystone cut.

very little effort, no matter how much they weigh when not having a powerful electric charge placed on them.

In a sense, this is the inverse process of piezoelectricity, which is a charge that accumulates in certain solid materials, notably crystals, when they are put under applied mechanical stress. Essentially it is electricity that is generated by putting pressure on a quartz crystal. Less understood is the inverse of this process, where an applied electric field causes the internal generation of a mechanical strain! In other words, when the crystal is electrified, it contracts or expands—it "bends" and flexes, changing its shape. This sudden change in the crystal causes the rectangular granite-crystal block to "jump." It also becomes briefly weightless during this period and could be "pushed" forward. It is an amazing concept! Sometimes the effect of pumping high voltage electricity into a crystal is called "the Hutchinson Effect," which refers to the Canadian experimenter John Hutchinson who is known for his videos on YouTube of objects losing weight and "floating" while under his "effect."

Picture this: huge granite blocks that have been quarried and dressed are then "electrified." This causes the block of crystalline stone to "bend" which causes it to become weightless. The blocks are then moved effortlessly through the air with guidelines such as ropes, or perhaps more high tech "pusher" beams of energy. A familiar movie scene eerily similar to this scenario is when the bounty-hunter Boba Fett in the *Star Wars* film *The Empire Strikes Back* takes the block of "carbonite" holding Han Solo aboard his ship. He is seen to be effortlessly pushing it ahead of him up a ramp onto his spacecraft. Could such a scene have been witnessed in ancient Peru? It seems incredible!

That the Incas actually found these megalithic ruins and then built on top of them, claiming them as their own, is not a particularly alarming theory. In fact, it is most probably the truth.

If the Incas came along and found walls and basic foundations of cities already in existence, why not just move in? Even today, all one needs to do is a little repair work and add a roof on some of the structures to make them habitable. Indeed, there is considerable evidence that the Incas merely found the structures and added to them. There are numerous legends that exist in the Andes that Sacsayhuaman, Machu Picchu, Tiahuanaco, and other megalithic remains were built by a race of giants.

Alain Gheerbrant comments in his footnotes to de la Vega's book:

> Three kinds of stone were used to build the fortress of Sacsayhuaman. Two of them, including those which provided the gigantic blocks for the outer wall, were found practically on the spot. Only the third kind of stone (black andesite), for the inside buildings, was brought from relatively distant quarries; the nearest quarries of black andesite were at Huaccoto and Rumicolca, nine and twenty-two miles from Cuzco respectively.
>
> With regard to the giant blocks of the outer wall, there is nothing to prove that they were not simply hewn from a mass of stone existing on the spot; this would solve the mystery. [40]

Gheerbrant is close in thinking that the Incas never moved those gigantic blocks in place, yet even if they did cut and dress the stones on the spot, fitting them together so perfectly would still require what modern engineers would call superhuman effort. Furthermore, the gigantic city of Tiahuanaco in Bolivia is similarly hewn from 100-ton blocks of stone. The quarries are many miles away, and the site is definitely of pre-Inca origin. Proponents of the theory that the Incas found these cities in the mountains and inhabited them would then say that the builders of Tiahuanaco, Sacsayhuaman, Ollantaytambo and megalithic structures in the Cuzco area were the same people.

Garcilaso de la Vega said this about these structures just after the conquest:

> ...how can we explain the fact that these Peruvian Indians were able to split, carve, lift, carry, hoist, and lower such enormous blocks of stone, which are more like pieces of a mountain than building stones, and that they accomplished this, as I said before, without the help of a single machine or instrument? An enigma such as this one cannot be easily solved without seeking the help of magic, particularly when

Stonehenge as viewed from the air.

one recalls the great familiarity of these people with devils.[40]

The Spanish dismantled as much of Sacsayhuaman as they could. When Cuzco was first conquered, Sacsayhuaman had three round towers at the top of the fortress, behind three concentric megalithic walls. These were taken apart stone by stone, and the stones used to build new structures for the Spanish.

One last intriguing observation which Protzen makes is that the cutting marks found on some of the stones are very similar to those found on the pyramidion of an unfinished obelisk at Aswan in Egypt. Is this a coincidence, or was there an ancient civilization with links to both sites?

The World's Largest Computer

The magnificent monument in England called Stonehenge sits alone on the Salisbury Plain, flanked by a parking lot and gift shop for tourists. It is famous for its large stones and curious architecture:

83

a circle of massive, well-cut stones.

In 1964 the British astronomer Gerald S. Hawkins first published his now-famous treatise on Stonehenge as an astronomical computer. His article, entitled "Stonehenge: A Neolithic Computer," appeared in issue 202 of the prestigious British journal *Nature*. In 1965, Hawkins' famous book *Stonehenge Decoded* was published.[28]

Hawkins upset the archaeological world by claiming that the megalithic site was not just a circular temple erected by some egocentric kings, but rather a sophisticated computer for observing the heavens.

He begins his *Nature* article with a quote from Diodorus on prehistoric Britain from his *History of the Ancient World*, written about 50 BC:

> The Moon as viewed from this island appears to be but a little distance from the Earth and to have on it prominences like those of the Earth, which are visible to the eye. The account is also given that the god [Moon?] visits the island every 19 years, the period in which the return of the stars to the same place in the heavens is accomplished... There is also on the island, both a magnificent sacred precinct of Apollo [Sun] and a notable temple... and the supervisors are called Boreadae, and succession to these positions is always kept in their family.

Hawkins' basic theory was that "Stonehenge was an observatory; the impartial mathematics of probability and the celestial sphere are on my side." Hawkins' first contention was that alignments between pairs of stones and other features, calculated with a computer from

Stonehenge restored to its original configuration by an artist.

small-scale plans, compared their directions with the azimuths of the rising and setting sun and moon, at the solstices and equinoxes, calculated for 1500 BC. Hawkins claimed to have found thirty-two "significant" alignments.

His second contention was that the fifty-six Aubrey holes were used as a "computer" (that is, as tally marks) for predicting movements of the moon and eclipses, for which he claims to have established a "hitherto unrecognized 56-year cycle with 15 percent irregularity; and that the rising of the full moon nearest the winter solstice over the Heel Stone always successfully predicted an eclipse. It is interesting to note that no more than half these eclipses were visible from Stonehenge."[28]

Says Hawkins in *Stonehenge Decoded*:

> The number 56 is of great significance for Stonehenge because it is the number of Aubrey holes set around the outer circle. Viewed from the center these holes are placed at equal spacings of azimuth around the horizon and therefore, they cannot mark the Sun, Moon or any celestial object. This is confirmed by the archaeologist's evidence; the holes have held fires and cremations of bodies, but have never held stones. Now, if the Stonehenge people desired to divide up the circle why did they not make 64 holes simply by bisecting segments of the circle—32, 16, 8, 4 and 2? I believe that the Aubrey holes provided a system for counting the years, one hole for each year, to aid in predicting the movement of the Moon. Perhaps cremations were performed in a particular Aubrey hole during the course of the year, or perhaps the hole was marked by a movable stone.
>
> Stonehenge can be used as a digital computing machine... The stones at hole 56 predict the year when an eclipse of the Sun or Moon will occur within 15 days of midwinter—the month of the winter Moon. It will also predict eclipses for the summer Moon.[28]

The critics of Hawkins, the ruling academic minds of their time, immediately jumped on his discoveries and denounced them. In 1966 an article by the British astronomer R. J. Atkinson appeared in *Nature* (volume 210, 1966), entitled "Decoder Misled?" in which Atkinson criticized Hawkins for many of his statements about Stonehenge being an astronomical computer.

Said Atkinson of Hawkins' book *Stonehenge Decoded*:

> It is tendentious, arrogant, slipshod and unconvincing, and does little to advance our understanding of Stonehenge.
>
> The first five chapters, on the legendary and archaeological background, have been uncritically compiled, and contain a number of bizarre interpretations and errors. The rest of the book is an unsuccessful attempt to substantiate the author's claim that 'Stonehenge was an observatory; the impartial mathematics of probability and the celestial sphere are on my side.' Of his two main contentions, the first concerns alignments between pairs of stones and other features, calculated with a computer from small-scale plans ill-adapted for this purpose.

Atkinson's scathing criticism of Hawkins is revealing because it shows how resistant to new ideas established academics can be. Atkinson's reluctance to believe that Stonehenge was some sort of astronomical computer is probably largely due to the popular belief that ancient man simply didn't have a state of civilization that allowed him to pursue such topics of higher knowledge.

But these critics are heard from no more, and there seems little doubt to even the most conservative archaeologist that Stonehenge is some sort of astronomical temple. There are a number of simple astronomical truths that can be discerned from Stonehenge. For instance, there are 29.53 days between full moons and there are 29 and a half monoliths in the outer Sarsen Circle.

There are 19 of the huge 'Blue Stones' in the inner horseshoe which has several possible explanations and uses. There are nearly 19 years between the extreme rising and setting points of the moon. Also, if a full moon occurs on a particular day of the year, say on the summer solstice, it is 19 years before another full moon occurs on the same day of the year. Finally, there are 19 eclipse years (or 223 full moons) between similar eclipses, such as an eclipse that occurs when the sun, moon and earth return to their same relative positions.

Other planets' positions may vary in even larger cycles.

It is also suggested that the five large trilithon archways represent the five planets visible to the naked eye: Mercury, Venus, Mars, Jupiter and Saturn.

The British writer on antiquities, John Ivimy, makes a stirring suggestion as the end of his popular book on Stonehenge, *The Sphinx and the Megaliths*.[76] He spends the bulk of the book trying to prove his thesis that Stonehenge was built by adventurous Egyptians who were sent to the British Isles to establish a series of astronomical sites at higher latitudes in order to accurately predict solar eclipses, which the observatories in Egypt could not do, because they were too close to the equator.

Ivimy gives such evidence as the megalithic construction, keystone cuts in the gigantic blocks of stone, the obvious astronomical purpose, and most of all, the use of a numbering system that is based on the number six, rather than the number ten, as we use today. Ivimy shows that the Egyptians used a numbering system based on the number six, and that Stonehenge was built using the same system.

It is a fascinating idea that Egyptians came to Britain to build a megalithic observatory to accurately predict lunar eclipses. It is recorded that in about 2000 BC a Chinese emperor put to death his two chief astronomers for failing to predict an eclipse of the sun. Asks ancient astronaut theorist Raymond Drake, "Today, would any king care?"

The Egyptians, Chinese, Mayans and many other ancient cultures were obsessed with eclipses as well as other planetary-solar phenomena. It is believed that they associated catastrophes, including the sinking of Atlantis, with planetary movements and eclipses. Perhaps the ancient Egyptians, Mayans and other civilizations thought they could predict the next cataclysm by monitoring lunar eclipses and the positions of the planets in relation to the earth.

Herodotus in his *The Histories* (written circa 440 BC) writes about ancient Egyptian astronomy and cataclysms in *Book Two*, chapter 142:

...Thus far the Egyptians and their priests told the story.

And they showed that there had been three hundred and forty-one generations of men from the first king unto this last, the priest of Hephaestus. ...Now in all this time, 11,340 years, they said that the sun had removed from his proper course four times; and had risen where he now sets, and set where he now rises; but nothing in Egypt was altered thereby, neither as touching the river nor as touching the fruits of the earth, nor concerning sicknesses or deaths.

If Herodotus is to be believed, Egyptian history goes back 11,340 years from 440 BC to the astonishing date of 11,780 BC. He is also saying that the earth has shifted around its axis in what is called a pole shift. The sun then appears to rise in a different direction from normal. Pole shifts are accompanied by a wide variety of devastating earth changes and severe weather phenomena. Therefore, if the Egyptians were familiar with this sort of occurrence, and having been thus far unaffected by the cataclysm, they may well have gone to great length to improve their astronomical knowledge, including the colonization of England and the building of Stonehenge.

The Egyptian Labyrinth of Herodotus

Herodotus also mentions a huge, maze-like structure that is apparently the largest structure in the world, located in sands to west of the Pyramids of Giza. He called the gigantic structure of over three thousand rooms "the Labyrinth" in *Book II* of his *Histories*, and describes it as a building complex in Egypt, "situated a little

The crumbling Pyramid of Hawara near the mysterious Labyrinth.

above the lake of Moeris and nearly opposite to that which is called the City of Crocodiles" that he considered to surpass the pyramids:

> It has twelve covered courts—six in a row facing north, six south—the gates of the one range exactly fronting the gates of the other. Inside, the building is of two stories and contains three thousand rooms, of which half are underground, and the other half directly above them. I was taken through the rooms in the upper story, so what I shall say of them is from my own observation, but the underground ones I can speak of only from report, because the Egyptians in charge refused to let me see them, as they contain the tombs of the kings who built the labyrinth, and also the tombs of the sacred crocodiles. The upper rooms, on the contrary, I did actually see, and it is hard to believe that they are the work of men; the baffling and intricate passages from room to room and from court to court were an endless wonder to me, as we passed from a courtyard into rooms, from rooms into galleries, from galleries into more rooms and thence into yet more courtyards. The roof of every chamber, courtyard, and gallery is, like the walls, of stone. The walls are covered with carved figures, and each court is exquisitely built of white marble and surrounded by a colonnade.

The Faiyum Oasis is a natural depression in the desert immediately to the west of the Nile south of Cairo. This large basin floor comprises fields that are watered by an artificial canal of the Nile created about 2300 BC, the Bahr Yussef, as it drains into Faiyum desert depression. The artificial Bahr Yussef canal veers west through a narrow neck of land between the archaeological sites of El Lahun and Gurob near Hawara and then branches out into the basin providing rich agricultural land. The water of the Faiyum basin (oasis) then drains into the large Lake Moeris. Lake Moeris was a freshwater lake in ancient Egyptian times but is today a saltwater lake. The lake was largely abandoned starting in 230 BC due to the nearest branch of the Nile dwindling in size and not having enough water to flood the oasis and the lake every year.

The artificial canal to the Faiyum Oasis served several purposes: control the Nile when it was flooding, regulate the water level of the Nile during dry seasons, and serve the surrounding area of Faiyum with irrigation. It is thought that the Egyptian pharaohs of the twelfth dynasty used the natural lake of Faiyum as a reservoir to

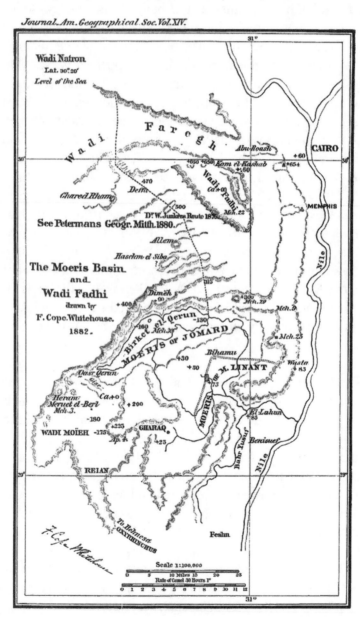

A map of Lake Moeris and region around the Labyrinth.

store surpluses of water for use during the dry periods. The immense waterworks of carving natural cliffs into the walls of the reservoir by the pharaohs of the twelfth dynasty, according to record, gave the impression that the lake and its stone walls was an artificial excavation, as reported by classic geographers and early travelers.

On the north side of the Bahr Yussef canal as it enters the Faiyum basin is the important archeological site of Hawara that contains the curious pyramid of Amenemhat III. Just north of Hawara was the ancient City of Crocodiles (Crocodilopolis). It was just south of this City of Crocodiles that the labyrinth was located according to Herodotus.

In 1889 the British Egyptologist Flinders Petrie (1853–1942) was doing archeological digs at the pyramid of Amenemhat III at Hawara. Flinders Petrie was a pioneer in systematic methodology within archaeology and preservation of artifacts during a dig. He excavated many of the most important Egyptian archaeological sites during the late 1800s in conjunction with his wife, Hilda. Petrie held the first chair of Egyptology in the United Kingdom and there is a Petrie Museum very near the British Museum in London which is famous for its many Egyptian artifacts, including solid stone core samples that Petrie maintained were evidence of sophisticated drill bits in ancient times.

During his 1889 archeological digs, Petrie began looking for the labyrinth described by Herodotus and others. Nearby the pyramid Petrie said he discovered an enormous artificial stone plateau that measured 304 meters by 244 meters. Petrie interpreted this artificial

Sir Flinders Petrie at Abydos, Egypt, in 1922.

91

plateau as being the foundation of the labyrinth. Petrie concluded that the labyrinth itself must have been destroyed at some time in antiquity and that that the only thing that was left was the stone base. Wikipedia says that in 1898, the *Harpers Dictionary of Classical Antiquities* described the labyrinth of Herodotus as "the largest of all the temples of Egypt, the so-called Labyrinth, of which, however, only the foundation stones have been preserved."

However, it is now thought that Petrie was not looking at the foundation of the labyrinth but rather at its ceiling. After all, one of the main features of the labyrinth was that the roof was made of huge slabs of granite. The first century BC Greek historian Diordorus Siculus said in his *History, Book I*, "this building had a roof made of a single stone" and another Greek historian, Strabo, said in his *Geography, Book 17*, "The wonder of it is the roofs of each chambers are made of single stones."

It would seem that the entire labyrinth—a massive granite building with hundreds of rooms—is actually underneath this feature that Petrie discovered and may occupy a much larger area than Petrie identified. Indeed, in 2008 an expedition led by a European researcher named Louis de Cordier—called the Mataha expedition—apparently got permission from the Supreme Council of Antiquities of Egypt, headed at the time by Dr. Zahi Hawass, for a team of geo-radar specialists from the National Research Institute of Astronomy and Geophysics to conduct extensive testing in the area identified by Petrie. The Mataha expedition and de Cordier said that the surveys revealed a strong suggestion of a vast number of chambers and walls several meters thick. Below the stone slab, at a depth of 8 to 12 meters, they found a "grid structure of gigantic size made of a very high

The crown of Lower Egypt.

resistivity material, such as granite."

De Cordier and the Mataha expedition believe that they have discovered the fabled labyrinth of antiquity, ignored since Petrie first thought he had discovered the labyrinth's foundation. De Cordier also believes that Dr. Hawass and the Supreme Council of Antiquities of Egypt are covering up the discoveries and currently no known excavations are taking place at this curious, megalithic structure.

What amazing artifacts might be found inside this subterranean labyrinth? What was the original purpose of this mysterious structure? It may be decades before we discover answers to these questions but one thing that we do know about the labyrinth—if it exists—it is one of the largest megalithic undertakings known to man and must incorporate millions of tons of hewn granite into its walls and ceilings.

The pyramid of Amenemhat III at nearby Hawara is also a curious structure. The pyramid is mainly constructed of millions of mud bricks but inside there are several passages that were to be cleverly blocked with 20-ton quartzite blocks built into special chambers that would later fall into place and seal a room when sand that was holding up the megalithic block was allowed to escape the passageway. Amenemhat III was the last powerful ruler of the 12th Dynasty, and the pyramid is believed to have been Amenemhet's final resting place, although his mummy was not found there. Nearby there is also a small pyramid tomb of Neferu-Ptah, Amenemhet III's daughter.

Petrie discovered no less than three quartzite 20-ton blocks within the passages but curiously none of them had ever been slid into place. He did not know whether this indicated negligence on the part of the burial party, an intention to return for more burials in the pyramid or a deliberate action to facilitate robbery of the tomb. Perhaps the purpose of the pyramid was completely different from being a tomb for the king, as they were often buried in mortuary temples near to the pyramids that they supposedly built, or in underground vaults such as at the Valley of Kings. The mummies of most pharaohs have never been found.

In what Petrie called the burial chamber he found that the floor was made of a single quartzite monolith that was lowered into a larger

An old drawing of the interior of the Serapeum.

chamber lined with limestone. Petrie estimated that this monolithic slab weighed about 110 tons. A course of brick was placed on the chamber to raise the ceiling and then a ceiling of three quartzite slabs weighing about 45 tons each was put in place. Above this chamber were two relieving chambers topped with 50-ton limestone slabs forming a massive pointed roof. Over these limestone slabs was an enormous arch of brick three feet high that was to support the core of the pyramid up to its summit.

Today the pyramid looks like a crumbling mass of mud bricks but inside of the huge structure are these megalithic slabs weighing up to 110 tons. Currently the entrance to the pyramid is flooded to a depth of six meters because the Bahr Yussef canal passes within 30 meters of the pyramid and has shifted course over the years.

The Astonishing Serapeum

A similar structure to the missing labyrinth is the underground vault near the Saqqara pyramids known as the Serapeum. This enigmatic structure and the huge basalt boxes inside it was closed by the Egyptian government for decades and finally reopened on September 20, 2012. Though I had visited the Serapeum once before, I was fortunate in 2007 to inspect the gigantic 70-ton basalt "sarcophagi" with engineer Christopher Dunn and his keen eye for telltale signs of power tools and ancient machining. But first, what is the Serapeum of Saqqara? We get a good definition from Wikipedia:

The Serapeum of Saqqara is a serapeum located north west of the Pyramid of Djoser at Saqqara, a necropolis near Memphis in Lower Egypt. It was the burial place of the Apis bulls, which were incarnations of the deity Ptah. It was believed that the bulls became immortal after death as Osiris Apis, shortened to Serapis in the Hellenic period. The most ancient burials found at this site date back to the reign of Amenhotep III.

In the 13th century BCE, Khaemweset, son of Ramesses II, ordered that a tunnel be excavated through one of the mountains, with side chambers designed to contain large granite sarcophagi weighing up to 70 tons each, which held the mummified remains of the bulls. A second tunnel, approximately 350 m in length, 5 m tall and 3 m wide (1,148.3×16.4×9.8 ft), was excavated under Psamtik I and later used by the Ptolemaic dynasty. The long boulevard leading to the ceremonial site, flanked by 600 sphinxes, was likely built under Nectanebo I.

The temple was discovered by Auguste Mariette, who had gone to Egypt to collect coptic manuscripts but later grew interested in the remains of the Saqqara necropolis. In 1850, Mariette found the head of one sphinx sticking out of the shifting desert sand dunes, cleared the sand, and

One of the basalt boxes inside the Serapeum.

followed the boulevard to the site. After using explosives to clear rocks blocking the entrance to the catacomb, he excavated most of the complex. Unfortunately, his notes of the excavation were lost, which has complicated the use of these burials in establishing Egyptian chronology. Mariette found one undisturbed burial, which is now at the Agricultural Museum in Cairo. The other 24 sarcophagi, of the bulls, had been robbed.

The French archeologist Auguste Mariette.

A controversial aspect of the Saqqara find is that for the period between the reign of Ramesses XI and the 23rd year of the reign of Osorkon II—about 250 years—only nine burials have been discovered, including three sarcophagi Mariette reported to have identified in a chamber too dangerous to excavate, which have not been located since. Because the average lifespan of a bull was between 25 and 28 years, Egyptologists believe that more burials should have been found. Furthermore, four of the burials attributed by Mariette to the reign of Ramesses XI have since been retrodated. Scholars who favor changes to the standard Egyptian chronology, such as David Rohl, have argued that the dating of the twentieth dynasty of Egypt should be pushed some 300 years later on the basis of the Saqqara discovery. Most scholars rebut that it is far more likely that some burials of sacred bulls are waiting to be discovered and excavated.

So, we see immediately that there are some unsolved mysteries in the Serapeum of Saqqara. The most obvious one is that only one mummy of the Apis bulls was ever found. All of the sarcophagi were empty except for one.

Egyptologists assume that the tomb was looted in antiquity and the mummified bulls were stolen. Why would anyone want to steal a mummified bull? Also, the archeologist Mariette had to use explosives to clear the way to the entrance to the underground complex—called a tomb—so it is clear that the Apis bulls have been missing for a very long time. Mariette also apparently found three other giant basalt boxes in another tunnel and that tunnel has not been found by subsequent archeologists. What? Is some undiscovered tunnel part of the current site? Who was Auguste Mariette?

French Archeologist Auguste Mariette

The French archeologist François Auguste Mariette was born at Boulogne-sur-Mer on February 11, 1821 and died in Cairo on January 19, 1881. He was a scholar and archaeologist, plus he is the founder of the Egyptian Department of Antiquities, now called the Supreme Council of Antiquities.

When his Egyptologist cousin died (Nestor L'Hote, who was the friend and fellow-traveler of Champollion), the task of sorting his papers gave Mariette a passion for Egyptology. He devoted himself to the study of hieroglyphics and the Coptic language. Mariette's 1847 analytic catalogue of the Egyptian Gallery of the Boulogne Museum got him a minor appointment in 1849 at the Louvre Museum in Paris.

A year later Mariette set out with a French government mission to seek and purchase the best Coptic, Syriac, Arabic and Ethiopic manuscripts for the Louvre collection, so it would retain its then-supremacy over any other national collections, such as the British or German. During 1850 Mariette tried to obtain manuscripts for the Louvre, but had little success. Since this might be his only trip to Egypt, he was determined to find something of archeological importance with which he could return to France. He befriended a Bedouin tribe while in the Giza area and the group led him by camel to Saqqara. Here he made the amazing discovery of the Serapeum. Says Wikipedia of Mariette's discovery:

> The site initially looked "a spectacle of desolation... [and] mounds of sand" (his words), but on noticing one sphinx from the reputed avenue of sphinxes, that led to the ruins of the Serapeum at Saqqara near the step-pyramid, with its head above the sands, he gathered 30 workmen. Thus, in 1851, he made his celebrated discovery of this avenue and eventually the subterranean tomb-temple complex of catacombs with their spectacular sarcophagi of the Apis bulls. Breaking through the rubble at the tomb entrance on November 12, he entered the complex, finding thousands of statues, bronze tablets and other treasures, but only one intact sarcophagus. He also found the virtually intact tomb of Prince Khaemweset, Ramesses II's son.
>
> Accused of theft and destruction by rival diggers and by the Egyptian authorities, Mariette began to rebury his finds in the desert to keep them from these competitors. Instead of manuscripts, official French funds were now advanced for the prosecution of his researches, and he remained in Egypt for four years, excavating, discovering—and dispatching

A photo showing the cramped spaces inside the Serapeum.

archaeological treasures to the Louvre, following the accepted Eurocentric convention. However, the French government and the Louvre set up an arrangement to divide the finds 50:50, so that upon his return to Paris 230 crates went to the Louvre (and he was raised to an assistant conservator), but an equal amount remained in Egypt.

Mariette returned to Egypt in less than a year and Isma'il Pasha (later to be called Ismail the Magnificent, Khedive of Egypt and Sudan from 1863 to 1879) created the position of conservator of Egyptian monuments for him in 1858. He moved his family to Cairo and was able to open the archeology museum in Cairo at Bulaq in 1863. He made excavations at Karnak in southern Egypt and even explored Gebel Barkal in Sudan. He also cleared the sands around the Sphinx down to the bare rock, and in doing so he discovered the famous megalithic granite monument, the "Temple of the Sphinx."

Mariette returned to France in 1867 to oversee the Parc Egyptien at the Exposition Universelle in Paris where he was welcomed as the world's preeminent Egyptologist. In 1869, at the request of the Egyptian Khedive, Mariette wrote a brief plot for an opera with Egyptian themes. Within the next year this plot was worked into a scenario by Camille du Locle, who proposed it to Giuseppe Verdi, who accepted it as a subject for his now-famous opera *Aida*, which opened in 1871.

European honors were bestowed on Mariette and in Egypt his was promoted to the rank of bey, and then pasha. The Cairo museum was ravaged by floods during the year 1878, and that destroyed most of his notes and drawings. In 1881, Mariette arranged for the appointment of the Frenchman Gaston Maspero to

Engineer Christopher Dunn examining basalt slabs at the Serapeum.

head the museum so as to ensure that France retained its supremacy in Egyptology, since the English now comprised the majority of archeologists in Egypt. Mariette died later that year in Cairo and was interred in a sarcophagus which is on display in a garden of the Egyptian Museum in Cairo. Some of his books include: *(Le) Sérapéum de Memphis*. Paris: Gide, 1857; *Catalogue général des monuments d'Abydos découverts pendant les fouilles de cette ville*. Paris: L'imprimerie nationale, 1880; *The Monuments of Upper Egypt*. Boston: H. Mansfield & J.W. Dearborn, 1890; *Outlines of Ancient Egyptian History*. New York: C. Scribner's Sons, 1892.

How Was the Serapeum Created?

Egyptologists seem to agree that the Serapeum, at least at a later period, was used in the worship of the Apis bull. Whether the gigantic basalt boxes, with lids weighing about 10 tons each, were originally built for some other purpose is another question. Also, just how were these 70-ton basalt vessels cut and drilled into? How were they moved through a small tunnel barely wide enough for the huge basalt blocks to squeeze through? Equally baffling is how these huge basalt blocks were maneuvered into the small bays that were cut for them out of the solid rock underground? The more one looks at the Serapeum the more incredible it seems!

In 2009 I was able to ask Christopher Dunn some questions about the moving of the huge basalt boxes, and he said that he had no idea how they would do it. He commented that he thought it would be a very difficult job considering the tight spaces and the tremendous weight of the boxes.

Our group had walked from the desert parking lot to the sloping ramp leading down to a locked steel door that covered the entrance. As we walked inside we came to a T-junction. Turning to the right we found the lid of a sarcophagus partially blocking the tunnel. Another tunnel met at this lid and some 20 feet down this tunnel is a huge sarcophagus stuck in the middle of the path. For unknown reasons the movement of this 70-ton basalt box was halted. Presumably it was on its way inside to take its place in an unoccupied niche made for the colossal stone block. It sits there in the tunnel as it has sat for thousands of years. The modern Egyptians working at the site certainly couldn't move the blocks, even if they wanted to. How the ancient Egyptians moved them is currently unexplained.

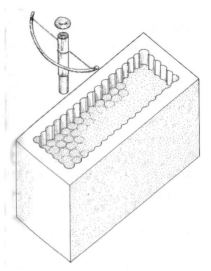

Moving to the left as one comes in from the entrance also brings one to a basalt lid partially blocking the tunnel. These pieces are critically heavy as well, and very difficult to budge, let alone place on top of the 8-foot high basalt boxes. Another corridor, called the main corridor, intersects some 20 feet beyond this lid. Some 22 basalt sarcophagi can be found along this corridor parked in special side bays that are cut out of the solid rock. The whole thing is about 50 feet underground, all carved out of solid granite. The quarries for the basalt must be some distance away.

As I walked around I wondered at the sheer mass of rock that would have been removed and the seemingly easy way that the ancient builders must have been able to move the basalt boxes. Did they have some sort of levitation device that made the 70-ton basalt blocks weightless so as to be easily moved through the tunnels? Had they somehow maneuvered them here with levers and rollers and sleds to eventually jack them into place? There simply wasn't space for more than a few people to stand around this block to push it into place, or the space for a block to turn at a right angle down a corridor. How was it done? The more I walked around the complex the more fantastic and sci-fi the whole scenario seemed. It was like the whole underground complex was impossible — some alien storage bunker.

The basalt sarcophagi themselves are amazingly smooth, straight and polished. Hieroglyphs are engraved on portions of the sarcophagi and these inscriptions are very fine on the hard polished basalt.

Chris Dunn took out some tools and measured the precision of the surfaces. The more precise they were the more likely that power tools were being used. The inside corners of the basalt boxes were especially difficult to make, Dunn said, and he thought that that would require power drills that drilled down again and again from above to eventually make a basalt box, one that could be used as a sarcophagus or as something else.

Dunn attempted to discover the radius of this drill bit—probably a diamond-tipped drill—by taking a measurement of the radius of one of the corners of the stone boxes. This radius he said would reveal the diameter of the drill that was being used to drill out the boxes (see illustration).

Dunn also felt that the boxes had to have been cut by computer-controlled machinery, like we have today, in order to make such perfect cuts. We looked at the delicate hieroglyphs that adorned some of the sides and they were perfectly cut. It was like some high-tech dentist's tool was etching the glyphs with the precision of a computer. There are also straight lines on some of the basalt sarcophagi and these lines are also of a highly precise nature.

In our discussions while inside the underground catacombs we all agreed that the site was baffling, with extremely heavy items being moved around in very tight spots. Plus, the precision of the basalt boxes and the difficulty in making them—even with power tools—boggled the mind. We would not attempt to make such boxes today, and if we did, they would be very expensive and then difficult and expensive to move. Was all this impossible effort undertaken because of a bull cult? Perhaps it was. What was the Apis bull cult anyway?

The Apis Bull Cult

The Apis bull was said to be the most important animal-god to the ancient Egyptians, at least in the Memphis-Nile Delta area. It was an early deity of the ancient Egyptians who had an annual festival around the Apis bull similar to the Djed festival in the fall to celebrate Osiris and the wheat harvest. Says Wikipedia about the Apis bull:

> In Egyptian mythology, Apis or Hapis (alternatively spelled Hapi-ankh) is a sacred bull worshipped in the Memphis region. "Apis served as an intermediary between humans and an all-powerful god (originally Ptah, later Osiris, then Atum)."
>
> Apis was the most important of all the sacred animals in Egypt, and, as with the others, its importance increased

as time went on. Greek and Roman authors have much to say about Apis, the marks by which the black bull-calf was recognized, the manner of his conception by a ray from heaven, his house at Memphis with court for disporting himself, the mode of prognostication from his actions, the mourning at his death, his costly burial, and the rejoicings throughout the country when a new Apis was found. Auguste Mariette's excavation of the Serapeum of Saqqara revealed the tombs of over sixty animals, ranging from the time of Amenhotep III to that of Ptolemy Alexander. At first, each animal was buried in a separate tomb with a chapel built above it.

The cult of the Apis bull started at the very beginning of Egyptian history, probably as a fertility god connected to grain and the herds. According to Manetho, his worship was instituted by Kaiechos (Nebra?) of the Second Dynasty. Apis is named on very early monuments, but little is known of the divine animal before the New Kingdom. Ceremonial burials of bulls indicate that ritual sacrifice was part of the worship of the early cow deities and a bull might represent a king who became a deity after death. He was entitled "the renewal of the life" of the Memphite god Ptah: but

A drawing of the sacred procession of the Apis bull in ancient Egypt.

after death he became Osorapis, i.e. the Osiris Apis, just as dead humans were assimilated to Osiris, the king of the underworld. This Osorapis was identified with Serapis of the Hellenistic period and may well be identical with him. Greek writers make the Apis an incarnation of Osiris, ignoring the connection with Ptah.

Apis was the most popular of the three great bull cults of ancient Egypt (the others being the bulls Mnevis and Buchis). The worship of the Apis bull was continued by the Greeks and after them by the Romans, and lasted until almost 400 CE.

So the Apis bull was a vague deity that was important, perhaps since pre-Dynastic times. Its festival was a seven-day party when they drank lots of Egyptian beer, ate lots of bread, and probably feasted on roasted beef as well as their typical diet of ducks and fish.

Yet, why the impossible-to-make basalt sarcophagus to hold a ritually killed and mummified bull? Just a simple granite cubical inside the rock-cut Serapeum would suffice to hold a mummified bull. Why make the sophisticated basalt box that would create so much more work? Is it possible that the basalt boxes had a different purpose when originally built? If so, then what was that purpose? Were they boxes meant to resonate at a certain frequency? We simply do not know. I should point out that a similar granite box lies inside the King's Chamber of the Great Pyramid. It has been speculated by Christopher Dunn and others that it is a box meant to resonate at a special frequency.

Whether the basalt boxes were meant to hold mummified bulls is really beside the point because no matter what the purpose of the basalt blocks is—for mummies or something else—the very fact that they exist is a serious problem for modern archeology. They were doing something in the past that we cannot explain today. The ancients were quarrying, dressing, drilling and moving some of the hardest and heaviest stones known to have been worked by man.

In the case of the gigantic stones at Baalbek, each weighing over 1,000 tons, or with the monolithic granite obelisks that weigh typically 500 tons, these massive monoliths were moved in an outdoor setting where the theoretical thousands of workmen would be standing somewhere with a rope harness around them. Others perhaps would be tugging on the ropes of some tower-crane-and-pulley system helping to raise the gigantic stone. But at the Serapeum the spaces are so tight that it is difficult to imagine more than a few people being involved in the movement of these heavy basalt blocks. How did they do it?

With the Serapeum, as with so much of ancient Egypt, all is not as it seems. Power tools and levitation would seem to have been used there. Many of the monuments in ancient Egypt were made with power tools claims Dunn in his books and videos, his most recent book being *Lost Technologies of Ancient Egypt*.[17]

The genius of the megalith builders and their use of 50-ton and 110-ton stone slabs for their building projects is remarkable. Obelisks weighed from 100 tons to 500 tons or more (the Unfinished Obelisk, to be discussed shortly, would have weighed over 1,000 tons). Like the many megalithic walls all over the world, were obelisks also more prevalent in the ancient world than we realize? Were obelisks once all over the world, on nearly every continent? Who were these megalith masterminds erecting stone circles, standing stones, obelisks, pyramids and platforms around our planet? Doesn't there have to be some connection behind all of these astonishing works of architects and builders?

Indeed, the megalithic masterminds virtually colonized the world, from Egypt to England to the Americas, India, Vietnam, Easter Island and Tonga. Megaliths exist in such remote places as Manchuria, the Philippines, Mongolia, Japan and islands in Indonesia. The megalithic masterminds were once everywhere. But above their giant temples, pyramids and walls what were the colossal obelisks for? Were they simple monuments to the sun or were they erected for something more?

Chapter 3

Mysteries of the Unfinished Obelisk

Weiler's Law:
Nothing is impossible for the person
who doesn't have to do it himself.

The Mystery of the Unfinished Obelisk

The standard definition of an obelisk is a monolithic stone monument whose four sides, which generally carry inscriptions, gently taper into a pyramidion at the top. These massive, pointed shafts of polished granite were often capped with gold. Their size undoubtedly made them extremely difficult to move and raise.

The real purpose of these obelisks, in my estimation, was to act as antennas for receiving and possibly for transmitting energy or other signals such as radio or homing beacons. The dynastic Egyptians, however, used them for different purposes, mainly as monuments. The dynastic Egyptians, who had lost the underlying science, sometimes moved various obelisks from their original sites, currently unknown, and placed them as monuments in temples at Karnak, Heliopolis and a few other places. At some sites, such as Karnak, it seems that the obelisks were already standing when the immense temple with its walls and massive columns was built around them.

Typically the dynastic Egyptians would erect two obelisks in front of temples. There are exceptions to this, such as Hatshepsut's obelisk in the temple of Karnak and the Unfinished Obelisk, which does not seem to have a twin.

However, without a doubt, the dynastic Egyptians carved and erected some of the smaller obelisks. The larger obelisks were probably already fallen, or perhaps still in place, and then reused during dynastic times. Whether the ancient Egyptians had the technology to quarry, transport and erect huge obelisks is still a very serious debate today. While huge, erect obelisks are a simple archeological fact, how they came to be where they are is a major mystery.

Still unexplained are the following questions:
1. How were obelisks quarried?
2. How were obelisks removed from the quarry and transported?
3. How were obelisks moved onto the ships and later moved off?
4. How were obelisks transported to the erection site?
5. How were the obelisks erected?
6. What was the purpose of obelisks since it was such a tremendous effort to do all of the above?

While reasonable guesses and logical theories have been advanced for some of these questions, the exact techniques and various mysterious elements remain, as we shall see. Explanations such as the use of stone balls, rollers, and hardened copper saws have their problems—read on!

How to Quarry An Obelisk
Today, quarrymen cut and carve granite using saws with diamond-edged blades and hardened steel chisels. Ultrasonic drilling is also used. Modern Egyptologists assume that the ancient Egyptian quarrymen and stonemasons didn't have these modern tools. How, then, did they quarry and cut such clean lines in their obelisks and other monumental statuary? The quarrying, transporting, and lifting of gigantic blocks of stone has been an enduring mystery, one that has brought forth many fascinating theories, from the very simple to the very complex.

Even the lifting of obelisks has posed a great mystery and mainstream archeologists have had to admit that they don't really know how the obelisks were put into place, nor exactly how they were quarried.

In March of 1999, the NOVA television documentary team went to Egypt to produce a special on obelisks entitled *Pharaoh's Obelisk*. The documentary team brought along an ancient-tools expert to find out how ancient Egyptians quarried huge pieces of granite for their obelisks. NOVA traveled to an ancient quarry in Aswan, located

440 miles south of Cairo. This is where the ancient Egyptians found many of the huge granite stones they used for their monuments and statues.

One of the most famous stones left behind is the Unfinished Obelisk, more than twice the size of any known obelisk ever raised. Quarrymen apparently abandoned the obelisk when fractures appeared in its sides. However, the stone, still attached to bedrock, gives important clues to how the ancients may have quarried granite.

In the NOVA documentary, archeologist Mark Lehner, a key member of the expedition, crouches in a granite trench that runs

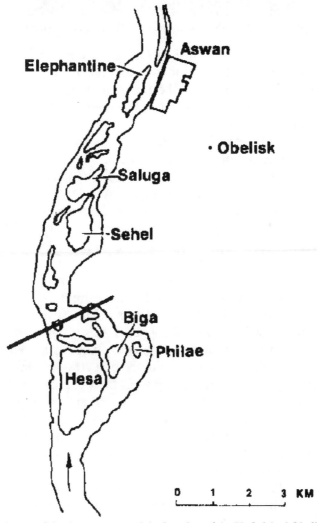

A map of the Aswan area and the location of the Unfinished Obelisk.

An early photo of the Unfinished Obelisk at the Aswan quarry.

along one side of the Unfinished Obelisk. Lehner holds a piece of dolerite similar to the kind that he and others believe Egyptian quarrymen used to pound out the trench around the edges of the obelisk. They presumably then lifted the pulverized granite dust out of the trenches with baskets.

It is theorized that workers pounded underneath the obelisk until the monument rested on a thin spine. Lehner says that huge levers were probably used to snap the obelisk from its spine, freeing it so it could be carved more finely and transported.

Archeologists know that the ancient Egyptians had the skills to forge bronze and copper tools. In the documentary, stonemason Roger Hopkins takes up a copper chisel, which works well when carving sandstone and limestone rock, to see if it might carve granite.

"We're losing a lot of metal and very little stone is falling off," observes Hopkins, which was hardly the desired result. Hopkins' simple experiment makes this much clear: The Egyptians needed better tools than soft bronze and copper chisels to carve granite.

NOVA then brought in Denys Stocks, who, as a young man, was obsessed with the Egyptians. For the past 20 years, this ancient-tools specialist had been recreating tools the Egyptians might have used. He believes Egyptians were able to cut and carve granite by adding a dash of one of Egypt's most common materials: sand.

"We're going to put sand inside the groove and we're going to put the saw on top of the sand," Stocks says. "Then we're going to let the sand do the cutting."

It does. The weight of the copper saw rubs the sand crystals, which are as hard as granite, against the stone. A groove soon appears in the granite. It's clear that this technique works well and could have been used by the ancient Egyptians.

But is this how the giant stone obelisks and other blocks were cut?

Hopkins' experience working with stone leads him to believe that one more ingredient, even more basic than sand, will improve the efficiency of the granite cutting: water. Water, Hopkins argues, will wash away dust that acts as a buffer to the sand, slowing the progress.

Adding water, though, makes it harder to pull the copper saw

An early photo of the Unfinished Obelisk at the Aswan quarry.

back and forth. While Hopkins is convinced water improves the speed of work, Stocks' measurements show that the rate of cutting is the same whether or not water is used.

Besides cutting clean surfaces on their granite, the Egyptians also drilled cylindrical holes into their stones. A hole eight inches in diameter was found drilled in a granite block at the Temple of Karnak. "Even with modern tools—stone chisels and diamond wheels—we would have a tough time doing such fine work in granite," says Hopkins.

Stocks was brought along to test his theories about how the cores were drilled. Inspired by a bow drill seen in an ancient Egyptian wall painting, Stocks designs a homemade bow drill. He wraps rope around a copper pipe that the Egyptians could have forged. Hopkins and Lehner then pull back and forth on the bow, which is weighted from above. The pipe spins in place, rubbing the sand, which etches a circle into the stone. With the assistance of the sand, the turning copper pipe succeeds in cutting a hole into the granite slab. But how can the drillers get the central core out?

In the NOVA documentary, Stocks wedges two chisels into the circular groove. The core breaks off at its base. Stocks reaches in and plucks it out, leaving a hole behind not unlike the ones once cut by the Egyptians.

This may well be how Egyptian workers were able to cut through granite with copper saws with the addition of a little abrasive sand. Sure, it was slow, hard work, but it could be done. Stocks' experiments squelched the various speculations about the ancient Egyptians possessing superhard tools or being able to soften stone. However, C. Ginenthal writes that H. Garland and C.O. Bannister attempted to saw through granite back in the 1920s using essentially the same method employed by Stocks—but without success. Garland and Bannister wrote a book on their experiments (*Ancient Egyptian Metallurgy*, London, 1927) on which Ginenthal comments:

> A consideration of the [copper and abrasive cutting] process would seem to give support to the idea that a copper-emery [or other abrasive material] process might have been used by the first Egyptians, but the author [Garland] has proved by experiment the impossibility of cutting granite or diorite by any similar means to these. ...no measurable progress could be made in the stone whilst the edge of the

copper blade wore away and was rendered useless, the bottom and sides of the groove being coated with particles of copper.

The Use and Reuse of Obelisks in Egyptian Times

Egypt as a civilization lasted for many thousands of years. Compared to our modern western civilization, only a few hundred years old, ancient Egypt has literally thousands of years of history. The beginnings of Egypt are said to go back to the "King Scorpion" (shades of the "Mummy" films) and the king known as Narmer. This time is generally thought to be around 3150 BC. Known as the Archaic Period, it is generally said to have lasted up to the Early Dynastic Period starting in 3050 BC.

The Early Dynastic Period began with the first "historical" pharoah, Menes, who is credited with "unifying" (at least symbolically) upper and lower Egypt. He was followed by the pharaohs Aha, Djer, Djet, and Den. The Early Dynastic Period ended around 2575 BC and the Old Kingdom began. The pyramids are thought by mainstream Egyptologists to have been built at this time.

The Middle Kingdom begins around 2040 BC and the New Kingdom, which included the reigns of Akhenaton, Nefertiti and Tutankhamun, began in 1550 BC, ending in 1070 BC. The ancient Egyptian culture basically came to an end in 343 BC with its conquest by Persia. Alexander the Great conquered Egypt in 332 BC, beginning the Greco-Roman period.

But what concerns us here is obelisks and their antiquity. Did obelisks exist in the land of Egypt before "King Scorpion" in 3150 BC? My contention is that they did indeed exist before that time, as did the Sphinx and the pyramids.

In fact, obelisks, along with various megalithic structures such as the pyramids, the Valley Temple of Chephren and the Osirion, may well be from a time many thousands of years before the currently accepted beginning of dynastic and predynastic Egypt. The theory presented here is that these structures are over 10,000 years old and have been occupied and reused by the dynastic Egyptians.

The Mystery of the Unfinished Obelisk of Aswan

Aswan is the capital of Aswan governorate in southeastern Egypt. About 708 km (440 mi) southeast of Cairo, it is on the east bank of the Nile River about 9.6 km (6 mi) north of the First Cataract.

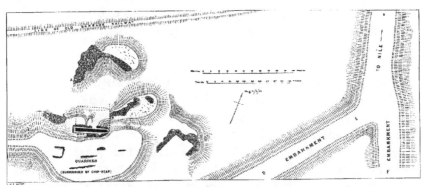

A map showing the Unfinished Obelisk in the quarry at Aswan.

The population swelled as a result of the construction (1960-70) of the Aswan High Dam. About 13 km (8 mi) south of the city, the dam has spurred development in the area.

The Aswan High Dam blocks the Nile River in Upper Egypt. One of the world's largest structures, the rock-fill dam, completed in 1970, has a volume about 17 times that of the Great Pyramid at Giza. It is 3.26 km (2.3 mi) in length and rises 111 m (364 ft) above the riverbed. Lake Nasser (High Dam Lake), the reservoir it impounds, averages 9.6 km (6 mi) wide and extends upstream 499 km (310 mi). About 30 percent of its length is in neighboring Sudan. An earlier granite dam, the Aswan Dam, lies 6.4 km (4 mi) downstream—about midway between the Aswan High Dam and the town. The Aswan Dam was completed in 1902, but its crest has been twice raised.

Ten years in construction, the Aswan High Dam cost $1 billion. The water it stores has opened the way to agricultural expansion. More than 360,000 hectares (900,000 acres), most of it formerly desert, were added to the total of arable land; an equal amount was irrigated year round to enable it to produce several crops a year instead of just one. Between 1979 and the mid-1980s, however, overuse and drought led to a 20% drop in the water level of Lake Nasser, forcing drastic reductions in the flow of irrigation water and reducing power output by 55%. The dam has a hydroelectric power capacity of 2.1 million kW and supplies more than 25% of Egypt's power.

Is it possible that an ancient dam at the First Cataract of the Nile also supplied power to an ancient power plant in the Aswan vicinity? Such a power plant, hydroelectric, with copper windings and cables, much as today's plants, could have existed circa 12,000

114

BC. A complete hydroelectric system may have been created—and is now gone. Disused metals, unless they are quickly looted (which is usually the case) will rust and oxidize within only a few hundred years. During this time modern-type quarrying was taking place (theoretically) at the quarries in Aswan. Giant obelisks were quarried, including the cracked "Unfinished Obelisk."

Also quarried would have been the stone blocks for the Osirion at Abydos. This megalithic site, half underwater in a swamp, is thought by many Egyptologists to be over 10,000 years old and is built out of perfectly cut blocks of Aswan granite weighing 100 tons or more.

At the edge of Aswan's northern granite quarries, separated from the bedrock, but lying in place, is what would have been the largest obelisk in Egypt. Because it cracked before the quarriers could lift it from its place and transport it to the Nile, we are able to follow the details of the ancient quarriers' art.

Says John Anthony West, the Egyptologist author of the books *Serpent In the Sky*[10] and *A Traveller's Key to Ancient Egypt*,[11] concerning the Unfinished Obelisk:

> If completed, this obelisk would have been a single monolith of granite 137 feet (43 m) long and 14 feet (4.3 m) thick at its thick end. It would have weighed 1,168 tons. The labor involved seems almost unimaginable. Apart from a few astronaut enthusiasts, everyone agrees that the Egyptians achieved their impressive results with the

A photo of the Unfinished Obelisk looking toward its base.

simplest possible means. Rock was quarried out of its bed by drilling a series of holes. Wooden stakes were inserted into the holes and soaked with water, and the expansion of the wood cracked the block out of the bedrock along the prescribed lines. The rough blocks were then pounded smooth with balls of dolerite, a rock even harder than the granite. A number of these pounding balls have been found. Assuming they were not brought down in spaceships by the alleged astronauts, the working-up of the dolerite into pounding balls then poses its own problem, and there are no definitive solutions. In all likelihood, the Egyptians had some simple yet sophisticated method of working with extremely hard abrasives—carborundum or even ground gems. The Egyptians had no steel, and the rare iron (probably from meteorites) had a ritual not a practical purpose. The Egyptians could, however, temper copper to a hardness close to that of steel by some method we also have not been able to reproduce. It is thought likely that they set their copper saws and drills with teeth and bits of precious gems, but there is little concrete evidence.

Who commissioned the obelisk is not known, nor where it might have been intended to go. The obelisk cracked from unknown causes, perhaps along a faultline invisible from the surface, in the final stage of freeing it from the bedrock.

Says West:

> Trenches had been cut all around the monolith, and the next step would have been undercutting it, and propping it up as work progressed. At that point either the vast rough piece of stone would have been levered somehow out of its trench, or the entire Nileside wall of granite that now encloses it would have been cut away, and it would have been ready to begin its several mile trek to the river, and the long ride to its destination, which remains unknown, as is the pharaoh responsible for ordering it. It is thought that it may have been cut for Hatshepsut.

I think work on the Unfinished Obelisk began long before the time of Hatshepsut, and likewise the giant obelisk named for her at Karnak. I think, in fact, that these are some of the giant crystals of Atlantis that are still around, used over and over. Some are still lying unfinished in their ancient quarry—a quarry that may be over 10,000 years old. Some were already standing when recorded history began, like the Osirion, the Sphinx, and the Giza Pyramids. Some were moved and erected again by the dynastic Egyptians, often to be moved and erected again by the Romans.

As the original purpose of these towers became lost, worship of their "power" began. When possible, they were moved to new temples being built for them. Otherwise, they were just too huge to move and erect. As I stated, they were, I believe, leftover crystal towers from "Atlantis."

Christopher Dunn, in his book *Lost Technologies of Ancient Egypt,*[17] has an entire chapter entitled "In the Shadow of an Obelisk." In this chapter he discusses the deep incisions carved into some obelisks and the difficulty of making these without power tools. He also discusses the type of machinery necessary to carve out an obelisk. Says Dunn:

> There are no surviving tools or machines that can be shown to have produced this work. Those that survive are incapable of such accuracy, especially on an industrial scale. There are some controversial theories about how the pyramids of Egypt were built, but the accepted conventional theory of copper chisels and stone or wooden hammers simply does not hold up because such technology cannot reproduce the results we see. Further, because this answer does not suffice, it invites nonconventional solutions. Yet these are not heard by the general public, who know of only the conventional theories that all children are taught in school and that audiences see on the Discovery Channel and on PBS when the manufacturing methods of the ancient Egyptians are discussed. On such programs, copper chisels and stone pounders, crude as they may be, are not posited as a possible method of manufacturing; their existence in the archaeological record is presented as proof that such tools were used to produce the monumental megaliths that dot the Egyptian landscape. Obelisks are a prime example of what such a crude technology was supposed to have achieved.
>
> It is important to give as accurate a description as possible of the characteristics of the sunken reliefs in Egyptian obelisks in order to judge whether modern attempts to show

how they were created satisfy the evidence. For instance, in the PBS Nova documentary *Secrets of Lost Empires: Obelisks*, Roger Hopkins, a stonemason who participated in the making of this documentary as a consultant and expert witness, discusses the reliefs with Egyptologist Mark Lehner: "Even with modern tools, and you know, diamond wheels and all that, we would have, you know, we would have a tough time getting it to this kind of perfection."

Not deterred by Hopkins' expert opinion, Mark Lehner picks up a dolerite pounder and demonstrates his theory of how the ancient Egyptains roughed out big hieroglyphs using it. After pounding for an hour he sincerely declares: "I'm convinced that with their skill their rapport with the stone and a great deal of time and patience, that this is the way they carved the fine details like the hieroglyphs on the obelisk."

…To his credit, Lehner admits that his efforts fell short of the quality of the ancient work on the simplest of shapes: the symbol for Ra, or the sun. His efforts produced a very shallow and rough relief compared to the original smooth, perfect profiles that are incised almost 1 inch deep. If he had managed to sink a perfectly formed falcon with narrow cuts of 0.14 inch wide, he might be able to argue that… [this] is an accurate representation of how the ancient Egyptians may have performed such intricate carvings.[17]

Dunn then turns his attention to the excavation of the Unfinished Obelisk in Aswan. This massive piece of red granite, similar to the rhyolite that is found at the Ollantaytambo quarry in Peru, weighs (if it had been extracted) an estimated 1,168 tons. Dunn discusses the supposed reason for the abandoning of the work on the obelisk: a crack was discovered in the rock during the quarrying, and therefore

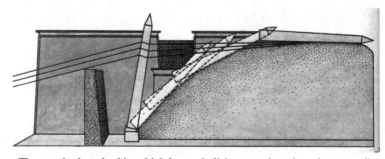

The standard method in which large obelisks were thought to be erected.

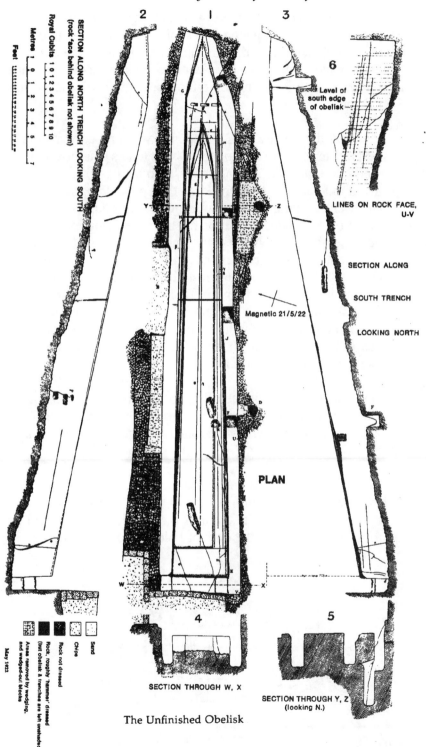

SECTION ALONG NORTH TRENCH LOOKING SOUTH
(rock face behind obelisk not shown)

Royal Cubits 10 1 2 3 4 5 6 7 8 9 10

Metres 1 0 1 2 3 4 5 6 7

Feet

2

1

3

6

Level of
south edge
of obelisk

LINES ON ROCK FACE,
U-V

SECTION ALONG

SOUTH TRENCH

LOOKING NORTH

Magnetic 21/5/22

PLAN

Sand

Chips

Rock not dressed

Rock, roughly 'hammer' dressed
(but obelisk & trenches are left unshaded)

Areas removed by wedging,
and wedged-out blocks

May 1922

W

X

Y

Z

4

SECTION THROUGH W, X

5

SECTION THROUGH Y, Z
(looking N.)

The Unfinished Obelisk

all work came to a halt. But Dunn wonders why so much effort would be expended on this gigantic block of granite and then suddenly halted—much like the work at the quarry at Ollantaytambo. He says that the obelisk could have been cut into smaller blocks to be used for statues or into larger building blocks, but this was never done. Says Dunn:

> Why the obelisk was abandoned will probably always be a mystery. There are no records that tell us that the quarry workers expended an enormous amount of resources on the granite and found a crack, so they picked up their tools and went to hammer somewhere else. We could speculate that all work ended while the obelisk was being excavated because of a cataclysmic event that disrupted the Egyptian civilization. All quarrying ended at Aswan and elsewhere, and it wasn't taken up again until the Romans controlled the country in the first century BCE. In fact, the Unfinished Obelisk may be the last artifact to be quarried in ancient Egypt, and as such, it would represent the state of the art in quarrying and moving large objects—notwithstanding the fact that it was not moved. [17]

Dunn says that he noticed that the Unfinished Obelisk had a deep trench, with unusual circular grooves visible in the trench, where the granite was removed around the obelisk. What he noted were vertical cuts into the rock with horizontal striations on the vertical cuts. This he ultimately surmised was the work of a drilling machine with an abrasive belt and drill that was a "megamachine" that plunged into the bedrock, cutting into it and removing rock. It would be constantly pulled out of the trench and reinserted into the rock, either to make the current hole deeper or to begin a new

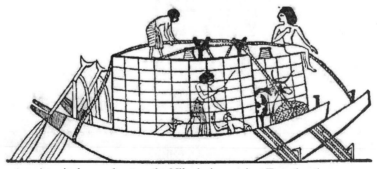

120   A typical cargo boat on the Nile during ancient Egyptian times.

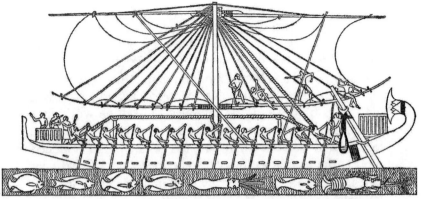

An ocean-going vessel making the trip to the mysterious land of Punt.

cut in the rock. This megamachine was a powered tool that was as advanced as anything we have today, Dunn claims, and something like it must have been used in the excavation of the obelisk.

Curiously, the Swiss stonemason Jean-Pierre Protzen says that work marks at the Ollantaytambo quarry in Peru are very similar to work marks that are found at the Unfinished Obelisk in Egypt. Were both made with the megamachines described by Dunn? Says Protzen:

> The work marks on the largest blocks of coarse-grained rhyolite are intriguing. They are found in three distinct patterns: roughly circular contiguous cups; approximately square-shaped adjoining pans; and adjacent parallel troughs. The cups are from 15 to 25 centimeters in diameter; the pans vary from 15 to 30 centimeters in width and 30 to 50 centimeters in length; and the troughs are from 15 to 50 centimeters wide. Cups, pans, and troughs are about 2 to 5 centimeters deep. Many of the large blocks in the quarries have a residual bulge along one or more of the bottom edges. The work marks stop at these bulges, which project from 10 to 30 centimeters.
>
> The stone cutting marks at *Kachiqhata* [Ollantaytambo] recall those found on the Unfinished Obelisk at Aswan. The Incas' cutting technique must not have been very different from the one used by the early Egyptians, who pounded away at the workpiece with balls of dolerite until it had the desired shape. Indeed, the Inca quarrymen and stonemasons did use hammerstones to cut and dress their building blocks.

121

The cups, pans, and troughs were the result of pounding or pecking at the workpiece with other stones. Working one's way down a block's face, one reached a point close to the bottom edge, where there was not enough room left between the bottom edge and the ground to pound effectively on the workpiece. Here, the stonecutter stopped, leaving a residual bulge. This bulge could be removed only if the workpiece was raised or turned over. On some blocks—for example, blocks 5 and 8 in the southern quarry—the bulge is indeed found at the top of the block. Since these blocks also bear work marks on their underside, it is reasonable to assume they were turned over for further work. [51]

It is extremely interesting that the marks at the *Kachiqhata* quarry are admitted to be very similar to those at Aswan. This telltale scooping and the cup marks that we can see are marks that Christopher Dunn thinks were made with a megamachine that was drilling into the rock. Protzen sees these cup and scoop marks as evidence of continued pounding and pounding with a stone hammer (usually a ball of rock about the size of a baseball) against the rock. He also says that the giant blocks of stone needed to be "turned over" so the ridges at the bottom of the block could be hammered away.

Yet, turning these blocks over seems to be a feat that may have been beyond the technology of the Incas. Since these blocks can weigh over 100 tons, one might think that it would be necessary to use levers to turn them over. Yet, Protzen has concluded that levers

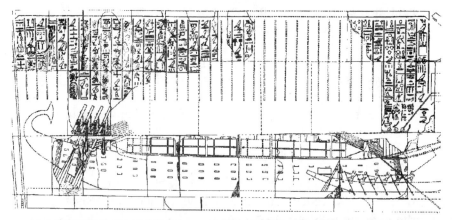

A depiction of an obelisk being transported on an Egyptian ship from El Bahari.

were not used by the Incas (or whoever was moving the blocks). How then was this done? Protzen gives no explanation. Indeed, it would seem impossible that levers were not used in turning over the blocks, which had to have been done. Why did they not use them in the arduous task of hauling the blocks, which, as we have seen, Protzen posits would have been incredibly laborious, with over a thousand people pulling on special harnesses to move the block even a few meters? Protzen's scenario is a strange mixture of crude and semi-advanced technology, all of which would require a tremendous amount of energy, manpower and man-hours of labor. Yet, he still cannot figure it all out, and concludes that it is all quite baffling and some things will just have to be left unexplained.

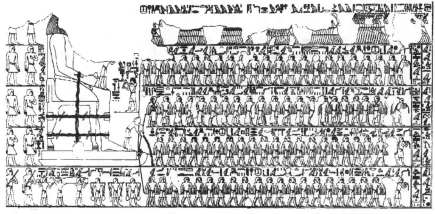

A depiction of a large statue being transported seen in the tomb of Djehutihotep.

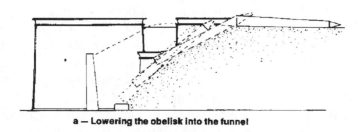

a — Lowering the obelisk into the funnel

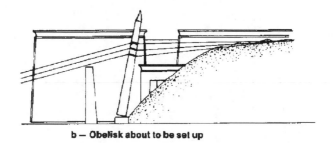

b — Obelisk about to be set up

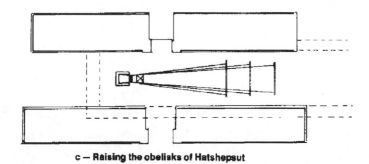

c — Raising the obelisks of Hatshepsut

The standard method in which large obelisks were thought to be erected.

Chapter 4

The Problem of the Obelisks

By definition, when you are investigating the unknown,
you do not know what you will find
or even when you have found it.
— *Bassagordian's Basic Principle and Ultimate Axiom*

In 1923 a study of obelisks was published entitled *The Problem of the Obelisks* by the British engineer and archeologist Reginald Engelbach.[8] The modern craze for all things Egyptological was in full swing and the public craved more and more information on ancient Egypt. And so it was inevitable that a book was written about that most mysterious (and heavy) of all ancient Egyptian objects: the obelisk.

Engelbach's now rare book was only about 150-pages but it included some good photos and diagrams and was the first book to look carefully at these puzzling monoliths. The book has been the bible for the few researchers who bothered to look into the various "problems" that are associated with quarrying, lifting, moving and erecting an obelisk—an endeavor that is major challenge for any engineer. The purpose of an obelisk is another problem, as the explanation of an obelisk is so vague in modern Egyptology as to be laughable. Engelbach says, in contrast to later explanations, that obelisks "had no very definite connection with sun-worship, their only function being an additional decoration to the pylons, though it is known that they were greatly venerated and offerings were made to them."

Says Engelbach in the complete quote from his first chapter, "Obelisks and Quarries" in which he tells us that some cities were filled with obelisks (where have they all gone?):

In ancient times there must have been a great number

125

of large obelisks in Egypt. Seti I tells us that he "filled Heliopolis with obelisks," and Ramesses II is known to have had fourteen in Tanis alone, though whether he erected them or merely usurped them, according to his habit, is uncertain. Besides the temples of the great centers such as Karnak and Luxor, Heliopolis and Tanis, many of the temples in other places must have had them. We have actual records of obelisks at Philae, Elephantine, Soleb (in Nubia), the mortuary temple of Amenophis III behind the Colossi of Thebes, and elsewhere. The total number of obelisks

The Unfinished Obelisk shortly after it was uncovered at the Aswan quarry.

exceeding 30 feet in length must have been well over fifty. The origin and religious significance of the obelisk are somewhat obscure.

In the royal sanctuaries of the fifth dynasty kings on the margin of the western desert at Abusir, not far from the Pyramids of Gizeh, the obelisk took the place of the holy of holies of the later temples. Recent excavations have shown that these obelisks were very different from those now familiar to visitors, as the length of the base was fully one-third that of the shaft, which was of masonry and merely served the purpose of elevating the sacred pyramid or benben, as the Egyptians called it—the real emblem of the sun. The obelisks of Upper Egypt, on the other hand, had no very definite connection with sun-worship, their only function being an additional decoration to the pylons, though it is known that they were greatly venerated and offerings were made to them. They were erected in pairs, and when Tuthmosis III put up a single one at Karnak, he says that it was the first time that this had been done.

Until we know how early obelisks were placed before the pylons of Upper Egypt, it is rather difficult to say whether they were developed from the fifth dynasty sun-obelisks or independently, particularly when one realizes that, if a high, thin stone monument is desired, the obelisk is the only practical form which is pleasing to the eye and convenient for inscribing. In any case, the subject is really outside the scope of this book, which deals rather with the mechanical side of obelisk-lore.

He goes on to tell us that the Unfinished Obelisk was first cleared of dirt and debris in 1922 and that some Egyptian obelisks had copper caps or even electrum (silver and gold) or pure gold. Gold was commonly mixed with other metals to create various alloys in Egypt, Sumeria, India and Peru. Southeast Asia was known in Sanskrit as Suvarnabumi, or the "Land of Gold" and it included Thailand, Cambodia, Vietnam, Malaysia, Indonesia and islands in the Pacific. Gold is a good conductor for all types of electricity and earth energies and will not rust, making artifacts that are made out of gold nearly indestructible. Therefore, adding a small amount of gold to any alloy will increase its durability as well as conductibility. However Engelbach sees such embellishment on obelisks as a mere craze for gilding objects. Says Engelbach:

> A discussion of the obelisk as a sun-emblem pure and simple is given in Prof. J. H. Breasted's *Development of Religion and Thought in Ancient Egypt* (Hodder and Stoughton) on pages 11, 15 and 71. The artistic taste of the ancient Egyptian differed considerably from ours and, to our minds, he was in the habit of decorating objects which do not need any decoration whatever. He had—like the modern Egyptian—a perfect mania for painting and gilding everything. In the tomb of Ramose at Thebes (No. 55) he has painted in gaudy colors the most wonderfully detailed reliefs, and we know for certain that he overlaid the huge fir-trees, which formed the pylon flagstaves, with bands and tips of electrum or copper. Obelisks did not escape this craze, and as far back as our records go they were capped with electrum, copper or gold. The Arab historian 'Abd El-Latif, writing as late as 1201 A.D., states that the two Heliopolis (Mataria) obelisks still retained their copper caps, and that around them were other obelisks large and small, too numerous to mention.
>
> Now only one remains. The unfinished obelisk of Aswan, though its existence has been known for centuries, was never cleared until the end of the winter of 1922, when my Department granted me L.E. 75 to do so. In this work I was assisted by Mahmud Eff. Mohammad and Mustafa Eff. Hasan of the Antiquities Department, who supervised the workmen. Before the clearance, all the visitor could see of the obelisk was the top surface of the pyramidion and about 20 yards of shaft, which sloped down into a vast heap of sand, chips and granite boulders. It has now become one of the most visited sights in Aswan, since nothing of its kind is to be seen elsewhere. Most persons, having seen the temples and tombs of Egypt, become more or less blasé to them. This is largely due to the fact that no one—least of all the dragomans—brings home to them the enormous difficulties the Egyptians overcame. They dismiss them as beyond their understanding, and many closer students of the monuments than the average visitor have boldly affirmed that the Egyptians knew engines and forces of nature of which we are to-day ignorant.

Engelbach acknowledges that some students of the monuments believed that the Egyptians knew of machines and forces of nature of which we are currently ignorant. At the time of his writing this text electricity and broadcast power, such as the towers that Tesla was working on, were in their infancy. Engelbach just couldn't know about the many functions of towers and antennas in the broadcast of radio, television, and telephone signals and even power for lamps and other electrical devices.

Engelbach feels that there is a step-by-step creation of an obelisk that is identifiable, at least in the actual cutting of the gigantic object from an outcropping of granite. On the difficulty of quarrying obelisks he says:

> Though modern research robs the Egyptians of the magical powers attributed to them, it makes them more admirable in the eyes of the practical man, as it shows that they could do, with the most primitive tools, feats of engineering which we, with some 3,000 years of mechanical progress behind us, are barely able to copy. A study of the Aswan Obelisk enables the visitor to look with different eyes on the finished monuments, and to realize, not only the immense labor expended in transporting the giant blocks and the years of tedious extraction of stone in the quarries, but the heartbreaking failures which must sometimes have driven the old engineers to the verge of despair before a perfect monument could be presented by the king to his god.
>
> Nowadays, if anything gets out of position, a jack, a winch or a crane is called for, and the trouble is soon put right; in ancient times a colossus or an obelisk which came down badly on to its pedestal was something in the nature of a tragedy.

Engelbach admires the workmanship of the stonemasons and waxes poetic on the broken coffin lids, unfinished sarcophagi and statues, abandoned pieces of granite or basalt vessels or objects that were mysteriously left unfinished in the outdoor workshop of granite blocks and chips near the Unfinished Obelisk:

> A walk round the quarries between the railway and the reservoir road at Aswan well repays the trouble. Here we

129

may see gigantic embankments, some nearly half a mile in length, on which the great blocks were transported from the high desert down to the Nile; we can see half-finished sarcophagi and statues, abandoned no one knows why, in various stages of completion; we can see inscriptions, some readable and some not, painted or cut on the boulders by the ancient engineers, and everywhere we may see the marks of their wedges, some showing where a block has been removed, others where the wedge has failed to act, or has split the rock in the wrong direction.

The site clamors for excavation, which might well reveal chippings from the chisels used in cutting the granite, and thus settle, once and for all, whether they were of highly tempered copper or not; another abandoned monument might give us conclusive information as to the methods by which they were detached from below, and how it was intended to roll them out from their beds. Excavation might well furnish us with ancient levers and rollers—or traces of them—which are hardly known at present, and then only of small size. A big quarry has never been cleared, and we cannot believe that the small area excavated round the obelisk has revealed all the secrets.

For thousands of years this large granite outcropping near the Egyptian city of Aswan, famous for the ancient temples at Elephantine Island, was essentially abandoned as a granite object manufacturing facility and allowed to be filled up with gravel, dirt and various bits of ancient trash until only the very summit, or tip, of the obelisk lying in the bedrock was visible. Now uncovered in 1922 archeologists and engineers were able to examine the giant object for the first time and were amazed at the awesome size of the monolith.

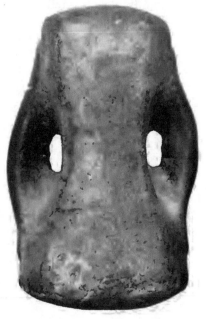

Photo of a black granite hammer from Giza.

130

Engelbach calmly acknowledges that the Unfinished Obelisk is "enormous" and then gives us the statistics, for the first time ever, of the unfinished monolith. He does not tell us that the typical obelisk weighs about 300 tons and a large obelisk weighs about 500 tons. Even a block of granite or basalt that weighed 50 to 80 tons would be a significant object to move. Engelbach's estimated weight of 1,168 tons if the obelisk had ever been extracted is discussed very matter-of-factly. That engineers from thousands of years ago were attempting to extract, move and erect such a massive piece of granite—for no discernable practical reason—must have left Engelbach with an awestruck wonder at the engineers and the monumental task in front of them.

Says Engelbach to the aspiring visitor to the site with its new designation as a tourist attraction for those visiting the Nile in the late 1920s:

> The best general view is obtained by passing over the new retaining wall at the butt, and thence up past the vertical face of rock to the hill above it. Even from there, owing to foreshortening, it is difficult to realize the enormous size of the monument, which is one-third as high again as the largest obelisk in Karnak, and more than triple the weight.
>
> Its complete dimensions are as follows:
> Length: 137 feet.
> Base: 13 feet 9 inches.
> Pyramidion base: 8 feet 2 inches.
> Pyramidion height: 14 feet 9 inches.
> Weight (if it had been extracted): 1,168 tons.
>
> As to the date of the obelisk, there is very little indication of it; since it was a failure, it was in nobody's interest to record it. It may have been of the time of Queen Hatshepsowet (i.e., about 1500 B.C.), since large obelisks seem to have been the rule in her time. Further, the outline of a smaller obelisk drawn upon the surface of the large one, which can be well seen just after sunrise, is of almost exactly the same dimensions as that now known as the Lateran obelisk at Rome, the work of Tuthmosis III, her co-regent and successor. These evidences of date should, however, be accepted with a good deal of caution.

The obelisk was abandoned owing to fissures in the granite, as the possibility of erecting a very large obelisk depends entirely on the rock being sound, particularly near the middle. Here, although the granite is of extremely good quality, it is by no means flawless, and from the very outset of the work the cracks and fissures seem to have given the ancient engineers a great deal of anxiety. Though parting fairly evenly under the action of wedges, the natural fissures in the granite are most erratic; a small fissure in one level or position may, in a couple of meters, become a gaping crack into which one could insert the blade of a knife; conversely, what appears to be a deep fissure may disappear at a lower level. Hence each crack had to be rigorously examined to see its probable effect on the completed obelisk.

Engelbach also briefly talks about the hundreds of stone balls of very hard dolerite—a granite substance that is harder than ordinary granite—that were used by at least some of the stonemasons. These dolerite balls are said to have been the main tool in smashing out large stone objects such as obelisks in granite quarries like the one at Aswan. Says Engelbach:

> ...All over the quarries at Aswan, and especially round

Engelbach's photo of the Unfinished Obelisk's trench in the quarry at Aswan.

132

the obelisk, may be seen hundreds of balls—some whole and some broken—of a very tough greenish-black stone known as dolerite, which occur naturally in some of the valleys in the eastern desert. It is a curious but incontestable fact that not only were the faces of monuments dressed by means of these balls—which has been long known—but that they were used for "cutting" out large monuments from the rock. In other words, they are the tools of the quarrymen.

On the setting out (or preparing it at the quarry) of the large Unfinished Obelisk at the Aswan quarry Engelbach says that they looked for cracks in the granite and then burned the top layer while pouring water on it to create cracks in the granite:

> At Aswan the surface of the granite consists of huge boulders, some quite large enough to provide a door-jamb or even a shrine, but none which could possibly furnish a moderate-sized obelisk.
>
> It must have required great experience to judge whether there was likely to be a long, flawless piece at a moderate depth. Whether test-shafts were sunk to examine the quality of the granite in all deep work I do not know, but I think it most probable, though in my superficial survey of the quarries I have not found any examples besides the two in the obelisk quarry. The quickest and most economical way of removing the top layers of the stratum is by burning fires against the rock, which causes it to break up very easily, especially if water is poured on it while it is still hot—a method used in India at the present day. There is a good deal of evidence to show that the Egyptians used this method, and it seems that the fires must have been of papyrus reeds, which at that time probably grew abundantly here just as it infests certain parts of the upper reaches of the Nile now. There are indications that these fires were banked with bricks against the surface to be destroyed. Traces of burning are seen at A and B and burnt granite can be picked up almost anywhere. It may be remarked here that the burnt granite must be distinguished from the weathered granite and that decomposed by the ferruginous layers in the stratum, which are likely to be confused with it. In the actual obelisk quarry, wedge-marks are seen only at one place.

133

The large blocks removed by a series of wedges acting in a channel instead of in slots are almost certainly of a later date than that of the obelisk. The (now) entrance to the trench is also a later piece of work, as the fine chisel-dressing is of the modern type, and I even obtained a block from here which had a hole "jumped" for blasting with gunpowder.

Engelbach then goes on about the stone balls and the controversial use of iron tools such as iron wedges on the granite at the Aswan quarry, something that modern Egyptologists discount (saying that Egyptians only had soft copper to work with):

I am inclined to think that the normal method was to use metal—perhaps iron—wedges, with thin metal plates between the wedge and the stone which are now known

Engelbach's photo of the interior of the trench in the quarry at Aswan.

as "feathers." The hammers may well have been of stone after the fashion of the Old Kingdom hammer from Gizeh (of black granite). The method used nowadays is to make, with a steel chisel, a series of small holes along the line where fracture is required, and by inserting small, fat, steel punches in them and giving them in turn, up and down the line, moderately hard blows with a sledge-hammer. In the clearance of the obelisk some hundreds of large blocks had to be broken up by this means before we could conveniently remove them. These had apparently been thrown down from the quarry above.

Ancient iron wedges, perhaps dating to 800 B.C., are given in Petrie, *Tools and Weapons*, Plate XIII, B 16, 17. Some enormous wedge-slots may be seen at the top of the rock in fig. 8, which may well have been cut for use with expanding wooden wedges. Having reduced the granite until they were satisfied that it was suitable for extracting an obelisk, and before dressing the surface in any way, they began to sink squarish holes round what was to be the perimeter or outline of the obelisk. This may well have been measured out by cords stretched over the rough surface. ... The method of making these pits is discussed in the next chapter. There are plenty of indications that they were begun before the surface of what was to be the obelisk had been made smooth.

For reasons which will appear later, the work on the pits progressed a good deal more slowly than that on the trench, so that, by the time the work had reached the stage at which it was abandoned, the trench workers had almost caught up with those engaged on the pits. Their object appears to have been to obtain as much knowledge of the state of the granite below as possible, especially as regards any horizontal fissures which might be met with, unsuspected from above.

The next step, an extremely laborious process, was to render the surface flat. This was done entirely by bruising with the balls of dolerite which have been found in such profusion in the quarry. Examples of unfinished top-dressing can be seen at the pyramidion and near the butt, where the work was abandoned early. Whether these balls were used by hand, or shod in some way on rammers, is doubtful. It seems likely that they were so mounted and worked by several men, as such blows were dealt that the balls were sometimes split in two—almost an impossibility by hand. A smooth straight surface along and across what was to be the upper face of the obelisk was almost certainly obtained by the use of what we now call "boning-rods."

These are a set of pieces of wood of exactly equal length, now usually made T-shaped. One rod is held upright at each end of the surface it is required to straighten. A man standing at one end can, if he sight along the top of these rods, see if a third rod, placed somewhere between them, is in a line with them or not. Thus the surface can be tested anywhere along the obelisk and corrected until it is quite flat. Boning-rods of small size, used for dressing moderately large blocks, have actually been found, and are published in Petrie, *Tools and Weapons*, Plate XLIX, B 44-46. These measure only about 3 inches high, and their tops were connected by a string. In the case of such a monument as an obelisk the string would sag and produce a concave error. The visual method, quite as simple and obvious, seems a legitimate assumption. The accuracy in the work of obelisks is not of a very high order, unlike the tremendous accuracy seen in the Pyramids of Gizeh and certain Old and Middle Kingdom monuments. An error in the sides of the base is quite usual, sometimes amounting to several inches.

Engelbach has said above that he believes that the dolerite balls that were found in abundance at the Unfinished Obelisk were used to render the surface to be flat by pounding with the stone balls—which are harder than granite—over and over again.

Engelbach then mentions how copper tools are too soft to do the work, but insists that the ancient Egyptians had iron and steel as the early British Egyptologist Flinders Petrie (previously mentioned), had also insisted. On the hardness of the tools that must have been used on the obelisks Engelbach says:

> An examination of the structure of ancient copper chisels shows conclusively that the copper had never been raised to the annealing temperature. It has been asserted that if the Egyptians had known steel it would have perished by oxydization. This is not borne out by excavations, as many iron tools have been found, such as wedges, halberds, etc.,

which are hardly rusted at all. In some soils almost anything will be preserved; in others everything, except perhaps the pottery, perishes. An examination of such fragments of iron tools as can be spared might give us some definite information as to whether any of them were of steel and so settle a vexed question. I have spent hours trying to cut granite with iron, copper, and even dolerite chisels, and though granite can be cut—in a manner of speaking—with all of them I am convinced that the Egyptians used a much harder tool. There is still a great divergence of opinion on this subject, which is best left open until further evidence is forthcoming.

On the extraction of an obelisk from the quarry he says:

The surface of the rock is smooth and the work on the pits around the obelisk is well under weigh. The next step seems to have been to mark on the surface of the rock the outline of the proposed obelisk. This must have been done by the normal Egyptian method of stretching a cord covered with ochre or lampblack over the proposed centre line and allowing the cord, when correctly placed, to touch the stone. The lines were next made permanent by scratching them with a metal tool. A pot containing red ochre was actually found during the clearance of the obelisk. The ochre or lampblack was probably mixed, before use, with acacia gum. From this centre line, by measuring off, the corners of the pyramidion and base were correctly marked and joined up. Let us examine the structure of the interior of the trench; we are struck with the absence of any marks of wedges or chisels. The ancient chisels leave traces which are easily recognizable, but here we have the effect of a series of parallel, vertical "cuts" just as if the rock had been extracted with a gigantic cheese-scoop. A further feature of the trench is that there are no corners—everything is rounded. These peculiarities are seen, not only in the trench, but in the pits within the trench and even the test-shafts C and D.

The only tools which could produce this effect are the dolerite balls of which we have already made mention. The trench and pits were therefore not cut out, but rather bashed out. These balls measure from 5 to 12 inches in

137

diameter, their weights averaging 12 pounds. They are of almost natural occurrence in some of the valleys in the eastern desert, having been shaped by the action of water in geological ages. A more economical or efficient tool can hardly be conceived. I have buried some hundreds of these behind the retaining wall, as even their size and weight did not protect them from souvenir-hunters.

The blows with these balls were struck vertically downwards, often with such force as to split them in two. This suggests that they were shod on to rammers, as it is almost impossible to break them by hand. The only way I succeeded in doing so was by pitching one down from a height on to a pile of others. This is further borne out by the fact that the wear on the balls is not even over the whole surface, but appears in patches, showing that they were used in one position until the bruising surface had become flat, and then changed to another position. If we enter the trench we see that, down the division between each concave "cut," a red line has been drawn, apparently by means of a plumb-bob with its string dipped in ochre.

On the pounding out of the trenches on either side of the obelisk he says that the powder created by the stone pounding balls had to be removed every few minutes:

When the granite is broken up by means of the dolerite balls or "pounders" it comes away in the form of powder and not as flakes. If the powder is not removed every few minutes, it soon forms a cushion, and the effect of the blows is reduced almost to nil. Handing the powder out of the trench would be a great waste of time, so it seems most likely that it was brushed on to the part of the task which was not being pounded; that is, each man worked on his task in four positions, with his back to, and facing, the obelisk on his right and left foot of trench. There is only one way by which such a large number of men can work in so confined a space without interfering with one another, and that is by making each man work in the same relative position on his task, and when a change in the position is required, by letting it be simultaneous.

...To return to the trench, it is interesting to speculate

on the amount of time which was expended in making it. To ascertain this, I tried pounding for an hour by hand at various times on one of the quarters of a two-foot task, and I found that I had reduced the level by about 5 millimeters (.2 inches) average. With practice I could perhaps have done more. Let us assume that the ancients could extract 8 millimeters .315 (.315 inches) per hour from a similar area; then the time taken to make the trench must be that taken to do the deepest part. In this obelisk the trench would have to be 165 inches to make it of square cross-section and we must allow at least 40 inches for under-cutting (p. 49), making a total depth of trench required of 205 inches. Supposing that .315 inches were extracted from a quarter of each party's task it will require 4x205 over 3.15x12x30 or 7.2 months of twelve hours per day. The undercutting would have taken at least as long again, even though it could be done from both sides at once.

On the tricky, final removal of the obelisk from the bedrock he says that wedges would be used to pry the gigantic monolith from the bedrock of granite:

> With regard to a large obelisk, I think we may safely say that it was neither snapped off its bed nor removed by the action of wedges from both sides. In a very long monument, the strains set up by the uneven expansions of the wedges, some biting true and some slipping out and not acting at all, would probably crack the monument in two, especially in the case of an obelisk like this, which could only safely stand the strains due to its own weight. It is fairly safe to assume that all large monuments were completely detached, perhaps by driving a series of galleries through first, packing them well by wood or stone as near the center of the monument as possible, and then removing the remainder of the rock. There is no evidence at all as to the nature of the packing.

On the transport of an obelisk weighing over a thousand tons Engelbach says the architect/engineer would have known what he was doing and believed that he could complete the monumental task:

It might be remarked that the Aswan obelisk—the largest known—has not been transported, but I think we are justified in assuming that the man responsible for the work would never have begun on it had he not every reason to believe that he could carry it out. Judging from such sketches as have come down to us of the character of Egyptian kings, they were not likely to tolerate a failure, unless it was from some unavoidable cause. We must bear in mind, too, that the ancient engineers moved blocks as heavy as this obelisk, and even more unmanageable—the colossi of Amenophis III and the colossus of Ramesses II at Thebes. We shall, therefore, take the Aswan obelisk as the basis of our speculations as, if we can account for every step in its history from the quarry to the temple, we can account for that of any other obelisk. The converse, reasoning from a small obelisk, would not necessarily be true.

The obelisk, then, is lying on its packing surrounded by the trench, but detached from the parent rock.

If we look at the surface of the rock outside the north (valley side) trench, we see that its level is the same as that of the surface of the obelisk. The parts A and B have most certainly been removed at a later date than the rest. It seems that a surface of rock, running continuously along the outside of the trench at the same level as that of the obelisk, was purposely left. It might be urged that this is merely the remainder of the flattened surface on which the obelisk was set out. This may well be the case, but if we consider in detail how the obelisk was to be gotten out of the pit in which it lies, factors arise which point to a very definite reason for leaving this surface as it now is.

There are two methods by which the obelisk can be removed from its present position: one is by raising it, and the other is by removing the rock from in front of it; sliding it out endways is impossible in this particular case. It may be mentioned here that to pull the obelisk over, on a level surface, would require some 13,000 men, which I am convinced could not be put on ropes in the constricted area of the quarry. To roll it out as it is would require an enormous quantity of rock to be removed, and one would think that, if they intended to use this method, they would have begun to do so as soon as possible. The fact remains, however, that they have not begun to do this, though they are well on with the breaking up of the rock to let the tip of the obelisk pass out.

A combination of both methods seems to have been intended, and the reason for leaving the north trench intact was for the use of large vertical levers. These would probably be tree-trunks, some two feet in diameter and 20 or more feet long, inserted, with suitable packing, in the trench, with many men pulling on ropes attached to the top of them. It seems that the workmen had begun to reduce the rock on the quarry side of the obelisk as well, so that levers could be used from there also. By using these levers from both sides of the obelisk in turn, it could be made to rock slightly backwards and forwards and gradually be raised by increasing the height of the packing below at each heave. By this means the base could be raised some 8 feet above its present level, and the quantity of rock to be removed from in

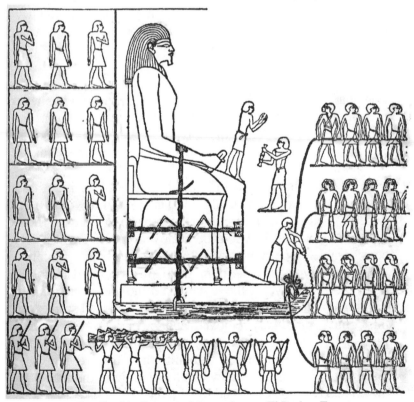

Engelbach's diagram of the tomb art at El Besher, Egypt.

front of the obelisk greatly reduced in consequence.

As to the numbers of levers needed; it can easily be calculated that, if they used thirty 20-foot tree-trunks at a leverage of six to one, with 50 men pulling on the ropes at the top of each, the obelisk would move, and the wood— whether it was of fir, cypress or sycamore-fig—would not be unduly strained. This is a conservative figure, and I think it likely that they would have used much taller trunks with at least 100 men pulling on each. On the further side of the obelisk, a comparatively small amount of rock would have to be removed in order to use the levers as, if they can move some 20 degrees back from the vertical, a sufficient rise in that side of the obelisk could be obtained.

As the base of the obelisk became higher, rock would have to be packed behind the levers, and on the valley side this would have to be very considerable, though with only 100 men per lever they could be used at a slope. As to the problem of packing the levers and keeping them steady, this is merely a matter of head-ropes and foot-ropes and could have been done in many ways. I do not propose to speculate on which particular method the Egyptians used, as there is no evidence on the subject.

Engelbach then describes how the engineers would use levers to raise the gigantic obelisk out of the bedrock that surrounded the massive tower of stone. This was only the very beginning of a long journey that is baffling to the modern scholar and engineer alike:

Directly after the obelisk had been raised as high as possible, the destruction of the rock in front of it would be done by wedging and burning, as described in Chapter III. I should think that it would be removed until there was a considerable slope downwards to the valley below, which would greatly reduce the number of men required to roll it. At the last heave of the levers from the valley side, the packing could be entirely withdrawn, and sand substituted; this could be gradually removed, and the obelisk allowed to settle down on to its edge and a great saving of men effected in this, its first and most difficult turn. By judiciously introducing a bank of sand where the middle of the face of the obelisk was to come, and by digging below its edge, the rolling could be made to approximate to that of a cylinder and its downward journey rendered comparatively easy.

The ropes for rolling the obelisk out would be passed round it and brought out to anchorages in front. I believe that 40 7-inch palm ropes (or their equivalent), pulled by 6,000 men, would be sufficient to handle the obelisk in any stage of its removal down the valley. Such large ropes would have to be pulled by handling-loops. In the scene of the transport of a great winged bull at Nineveh, they can be seen passing over the men's shoulders, being attached at both ends to the main cable. …

The occurrence of levers is so rare that it has been doubted whether the Egyptians knew of them. I think that there is not the slightest doubt that they did know of them, as in the temples of the Theban area and in the temple of the third pyramid at Gizeh, one can see large blocks, undercut at various points along their length, obviously to take the points of levers. In a tomb at El-Bersheh (*Annates du Service des Antiquitis*, I, p. 28), an acacia branch, with its end cut to a chisel edge, was found, which must have been used to manipulate the lid of the sarcophagus. It might be asked why no very large levers have been found. The reason is that large baulks would not be abandoned in the quarry, but would be used until they were no longer sound, and then cut up and re-used for other purposes. Like timber baulks today, they were of considerable value, and not thrown away when a job was completed. The Assyrians, at any rate, knew them, for in a sculpture of about the VIIIth century B.C. there is a scene of men hauling along a colossal bull mounted on a sled running on rollers, with men overcoming the initial friction with levers from behind (Layard, *Discoveries*, Plates X-XVII).

So Engelbach thinks that the use of levers, sleds and hundreds of men pulling on ropes can remove the obelisk and bring it to a boat waiting on the Nile. About the Queen Hatshepsowet sculpture at Der El-Bahari of obelisks being moved by sleds to a ship he comments:

We know, from the celebrated sculpture at Der El-Bahari, that the obelisks of Queen Hatshepsowet were

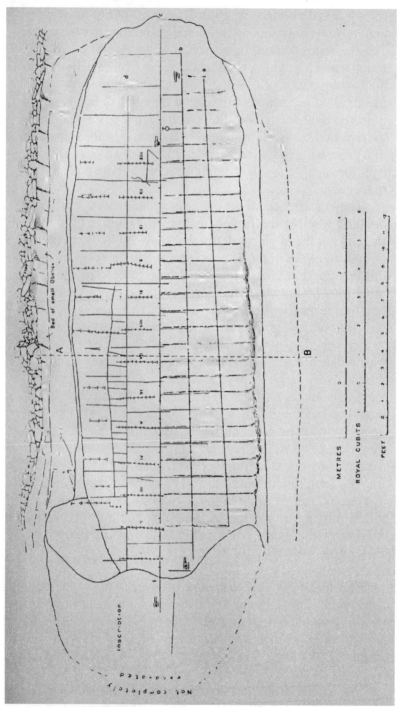

Engelbach's diagram of measuring lines on the quarry face at Aswan.

transported on sleds. It was probably done by the court artist from memory, and though the general impression is most likely correct, several of the details appear to be wrong. Thus he slurs over the manner in which the baulks of timber at the top of the obelisk were attached to those on the sled, which must have been done by the known Egyptian method of the "Spanish windlass," that is, by passing ropes round corresponding baulks and tightening them after the manner of a tourniquet (fig. 28, p. 70). The position of the hauling-rope in the center of the obelisk must also surely be wrong, as that would be the very worst position for pulling the obelisk; the rope would, of course, be attached to the sled, as it is shown in other similar scenes which have come down to us. It seems likely too that the obelisk was really on the sled the reverse way round (p. 70). The fact that the Der El-Bahari obelisks were mounted on sleds is no proof that all obelisks were so mounted for transport, but I think it most likely that they were, as without a sled it would be extremely difficult to attach ropes to the obelisk so as to be able to pull it lengthways; further, a sled would be an excellent shock-absorber and would equalize the upward pressure of the rollers along the length of the obelisk. This is almost a necessity in such a long obelisk as this, as, if it came down on a roller near its center with a jerk, it would snap in two.

He then tackles the problem of whether rollers would have been used to move the larger obelisks and he concludes that rollers must have been used, despite the doubts of other Egyptologists of the time. Incredibly, he suggests that calculations came up with a figure that 11,000 men would have been needed to pull the Unfinished Obelisk, something he says would have been simply impossible considering the limited space that they would have to stand in as they got close to the river. A canal built up to the obelisk might have been an easier task, but this was never done. Says Engelbach:

> Next comes the vexed question whether rollers were used in conjunction with the sled or not. It has been assumed by certain writers, because in the tomb scene of the transport of the 60-ton statue of Dhuthotpe at El-Bersheh (Lepsius, Denkmdler, II, 134, and p. 59) the sled was merely pulled over a wetted track, that all blocks were so transported,

whatever their size. When it is realized that it took 172 men—who would pull about 8 tons—to haul this statue, one hesitates to assert that a block of 1,170 tons was so handled. Caution is very necessary, but to deny that rollers were known in Egypt, as some writers would have us do, is either to invite far less justifiable assumptions, or to bring all reasoning to a standstill. The 227-ton obelisk now in Paris, when it was being pulled up a slight slope, mounted on a sled or "cradle" sliding over a greased way, required a pull of 94 tons. To handle the Aswan obelisk in this way would take at least 11,000 men, which is outside the bounds of possibility, if only from considerations of space. Small rollers have actually been found, but no large ones, for reasons already given for the absence of large levers. It is rather difficult to obtain data as to the size of rollers required for such an obelisk as that of Aswan. The only information I can give is that the top of the fallen obelisk of Queen Hapshepsowet rests on 8-inch diameter pitch-pine rollers, spaced about a yard apart, and there is no sign of any crushing, though they have been there for many years. The worst pressure at that spacing which they would have to bear at the butt-end of the Aswan obelisk, if it were placed on them for transport, would not exceed 11 times the amount they bear now.

The process of putting the obelisk on to its sled and rollers must have been something of this kind: at the foot of the slope leading down from the quarry the sled—mounted on its rollers and track baulks—would be buried, sighting-poles being put in to mark the position of its axis. The obelisk would then be rolled down the slope until it lay exactly over the sled, and the sand dug away till the obelisk settled down on to it. After digging the sled clear, the journey to the river could be begun, the track being packed as hard as possible, most probably with baulks of timber laid down lengthways on which the rollers could run. The route for the Aswan obelisk would almost certainly have been north-eastwards along the track of the old Barrage railway until it joined the embankment F-E which leads to the river. Its exact point of arrival at the Nile is hidden by the modern town.

Note that Engelbach says that sand was used in the loading of the obelisk onto the sled. Of the very few texts relating to obelisks

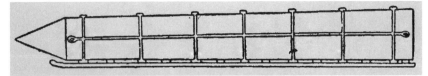

Engelbach's diagram of an obelisk attached to a sled for transport.

one of them does mention the removal of sand as part of the moving process.

Engelbach goes on to say that we do not know much about these boats and barges that were used to transport the obelisks. Furthermore, Engelbach sees the scenes as somewhat suspect on their accuracy:

> On the details of the enormous barges on which obelisks are known to have been transported, I have to be more than vague, as the only scene of a boat sufficiently large to carry an obelisk is that on the Der El-Bahari sculpture, where the two obelisks—probably those now seen at Karnak—are both placed butt to butt on the same barge! The boat used must have been over 200 feet in length. Another great barge is mentioned measuring 207 feet long by 69 feet broad, which carried the two obelisks of Tuthmosis I, and we have a record of a third boat in the Old Kingdom, made by one Uni of the VIth dynasty, which was 102 feet long, and which took only 17 days to build (Breasted, *Ancient Records*, I, 322, and II, 105).
>
> Mr. Somers Clarke, in *Ancient Egypt*, 1920, Parts 1 and 2 (Macmillan), has collected all known facts on the construction of ancient boats. He admits that the details of the very large ships are quite unknown, as the Der El-Bahari boat already referred to is only, as it were, an impressionist view, and from it we can learn little of its internal structure.
>
> The ancient boats in the Cairo Museum are only of quite small size; these are built without ribs, but whether the obelisk-barges were of this type also is uncertain. The Der El-Bahari boats are stiffened by means of a series of ropes attached to the bow and stern, passing over vertical supports at two points in the body of the boat, thus forming what is now known as a queen-truss or hog-frame. ...It is better to leave the question of the large boats until further evidence is forthcoming, but before doing so I will give a passage in

147

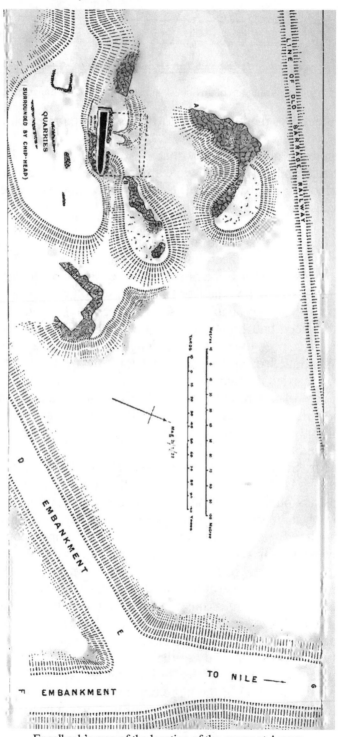

148   Engelbach's map of the location of the quarry at Aswan.

Mr. Clarke's article which is of interest to the general reader.

He says, quoting from a letter from the late Mr. Francis Elgar, Director of Naval Construction to the British Government: "The two great obelisks of Karnak, 97 feet 6 inches long, could be carried on a boat about 220 feet long and 69 feet beam, upon a draught of water of about 4 feet 6 inches or not exceeding 5 feet." Some of the large Cook's boats approach this length, but their beam is very different.

Mr. Clarke remarks later: "Whence came the necessary knowledge, at what period did the people begin to accumulate the experience, which culminated in their power to deal with immense weights... not only in the XIIth and XVIIIth dynasties, but in the IIIrd and IVth?"

On the gigantic barges that must have been used to tow the huge obelisks Engelbach says that there was only one way to load one with an obelisk. Again it requires the removal of sand or the river embankment:

It is a very great pity that the scenes of the transport by boat of Hatshepsowet's obelisks are not accompanied by a real descriptive text. All that we can learn from the inscriptions is that the boat was built of sycamore-fig, and the fact that a whole army was mustered at Elephantine, or Aswan, to load the obelisks on to it. There is plenty also about the rejoicings of the priests, marines and recruits over their arrival at Thebes. The scenes themselves, however, show us that the obelisk-barge was towed by three rows of oared tow-boats, which were arranged nine in a row, each row being led by a pilot-boat. Near the great barge are three boats escorting it, in which religious ceremonies are apparently being performed. We see the troops on the shore waiting to do the unloading, and an offering being performed by officials and priests. The name of the King, Tuthmosis III, is mentioned in the laudatory sentences after the Queen. In the view of the great barge, which is badly damaged, the obelisks are placed high up on her deck. This is possibly a trick by the artist so that they may be visible.

There is only one practical way of putting a large obelisk into a barge, and that is by getting the boat as close to the bank as possible, building an embankment round and over it, and pulling the obelisk directly over the boat and letting it down into its place by digging out the filling from beneath it. Possibly a new set of baulks and rollers were already prepared inside the barge. The boat would then be dug clear and the journey by water made. Though I see no reason to suppose that the rise and fall of the Nile were used for the loading and unloading of the boat, it is more than probable that it was arranged that the water journey was made at high Nile to minimize the risk of running aground.

The unloading would be a rather simpler matter. An embankment would be constructed from the shore to the boat (and around it), but only reaching to the level of the rollers of the obelisk. The boat would be destroyed—or at least the prow removed—and the journey continued towards the temple.

Chapter six of Engelbach's book is on the erection of obelisks, something that has been much debated and about which little is known. No Egyptian text or depiction of the raising of an obelisk exists. Engelbach says that modern engineers would have used some sort of hydraulic jack or screw to raise the obelisk but the Egyptians did not know of this device:

> The ancient method of setting up a large obelisk has been a fruitful subject for speculation for generations, and many extraordinary theories have been put forward by archaeologists, engineers, architects, and that bane of the serious student, the reckless exponent of the occult.
>
> In mediaeval and modern times, the erection of an obelisk has always involved capstans or winches actuating a system of pulleys, and in most cases a "jack"—either hydraulic or screw—has had to be called into use. It is generally admitted that the Egyptians were not familiar either with the screw-jack, capstan, winch or the system of pulleys arranged to give a mechanical advantage; it is even debatable whether they knew the simple pulley.
>
> Sheers were possibly known in principle, though we have no proof of it, but the erection of an obelisk by this means must involve the use of the capstan or winch. This leaves levers as the only source of power except the employment of large numbers of men. We have therefore to

try and explain how the large obelisks were erected by these means only.

Two theories stand out as being reasonable, though both leave a good deal unexplained. One is that the edge of the obelisk was placed so as to engage in the narrow notch which always runs along one side of the surface of the pedestals, and that it was gradually levered up, the earth being banked behind the levers at each heave, until the obelisk was leaning against an earth slope at a sufficiently steep angle to permit it to be easily pulled upright. This method was actually used for the erection of the memorial obelisk of Seringapatam, but the obelisk only weighed some 35 tons.

Some of the reasons against this having been the Egyptian method are as follows:

(a) The Egyptians could introduce obelisks inside courts whose walls were shorter than the length of the obelisk. Queen Hatshepsowet put hers between her father's pylons where there was a court of Osiris figures, and there is no evidence at all that any of the walls had been removed or rebuilt; in fact I am certain that they were not.

(b) Some obelisks are so close to their pylons that there would hardly be room for the huge levers which would have had to be used.

(c) After pulling the obelisk upright there is nothing to stop it from rocking about and getting out of control. The lowering of the New York obelisk showed clearly that, once it was on the move, head-ropes were more than un-reliable

Engelbach's photo of the gigantic embankment at the quarry at Aswan.

151

in checking the momentum of such a mass.

(d) The obelisk of Hatshepsowet at Karnak has come on to its pedestal askew, and has never used the notch at all, as its edge is quite sharp and unburred. This shows that the notch—an essential for this method—was not an essential for the ancient method.

Engelbach seems to back the second theory which involves the sliding of the obelisk down an embankment or a funnel-shaped pit to stand in a pit and then remove more sand from the obelisk. This technique however requires the obelisk to be raised to a considerable height above the ground where it was to stand. Says Engelbach:

> The other theory is that the obelisk was pulled up a long sloping embankment until it was at a height well above that of its balancing-point or "center of gravity," and that earth was cut from below it carefully until the obelisk settled down on to the pedestal with its edge in the pedestal-notch, leaning, as in the last method, against the end of the embankment. From thence it was pulled upright.
>
> The use of a large sloping embankment is more than likely, as (see note a above) the obelisk was obviously lowered on to its pedestal and not raised at all; this method,

Engelbach's sectional model of an embankment for an obelisk.

however, has some serious objections, which may be summed up briefly:

(a) It would be extremely risky business to cut earth from below an overhanging obelisk of 500 tons and upwards. Anyone who has seen earth undercut below a large stone in excavating work or elsewhere knows that the earth has a partiality for slipping sideways in any direction but the expected—preferably on to the heads of one's workmen.

(b) To make an obelisk settle down from a height on to a small pedestal by under-cutting would be an impossibility. Whatever method the Egyptians used, it was certain, and did not depend on the skill of the men with the pick and basket.

(c) See note c on the levering-up theory, which is equally applicable here.

A method which is mechanically possible and which meets all observed facts is that the obelisk was not let down over the edge of an embankment, but down a funnel-shaped pit in the end of it, the lowering being done by removing sand, with which the pit had been filled, from galleries leading into the bottom of it, and so allowing the obelisk to settle slowly down. Taking this as the basis of the method, the form of the pit resolves itself into a tapering square-sectioned funnel—rather like a petrol-funnel—fairly wide

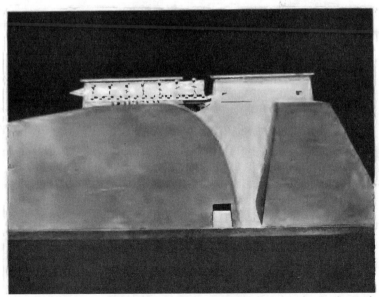

Engelbach's sectional model with the obelisk overhanging the sand tunnel.

153

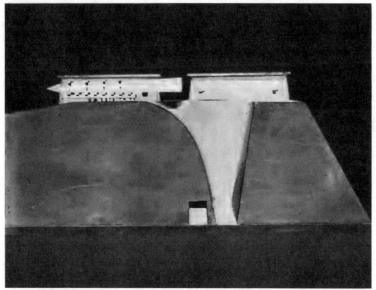

The model of the obelisk overhanging the tunnel with part of the sled removed.

at the top, but very little larger than the base of the obelisk at the bottom. The obelisk is introduced into the funnel on a curved way leading gradually from the surface of the embankment until it engages smoothly with the hither wall of the funnel. The sand is removed by men with baskets through galleries leading from the bottom of the funnel to convenient places outside the embankment.

He says that the obelisk had to come down the funnel—as workmen removed sand from beneath the monolith—to a granite platform which had a notch in it to help keep the obelisk from shifting from its desired placement on the pedestal:

> It is likely that part of the wall of the funnel had to be cut away to enable the obelisk to be pulled upright, though in any case I should imagine that enough space was left between the base of the obelisk and the funnel to enable men to get round and remove any stones, etc., which might have come down in the sand. It is seen, therefore, that the notch, although not an essential to the process of erection, is necessary for a perfect piece of work.
>
> As soon as the obelisk had come down into its notch, men would enter through the gallery leading in from the end

154

of the embankment, and clear every particle of sand from under the base, before it was pulled upright. Any tendency to rock after passing its dead-center could be avoided by filling the space between the obelisk and the further wall of the funnel with coarse brushwood to act as a sort of cushion. The reason why I suppose that the sand was removed from the front gallery (which leads into the left side of the funnel) is that, if it were removed through the end gallery, there would be a far greater likelihood for the obelisk to jam against the opposite wall, since the flow of sand would be forwards rather than from under the obelisk.

An alternative possibility for the form of the funnel is that it had vertical walls in a transverse sense, the width being but very slightly greater than that of the base of the obelisk; in other words, made so that the obelisk entered like a penny in the slot. By this means, full advantage would be taken of the weight of the sand above the obelisk, which would have the effect of bringing the base downward. I think, however, that the advantage gained would be discounted by the difficulty of controlling the descent by digging, etc., but it is a possibility which must be taken into serious consideration.

In the base of the now fallen obelisk of Tuthmosis III, which stood before the pylons of Tuthmosis I at Karnak,

Engelbach's sectional model with the obelisk now in the sand tunnel.

155

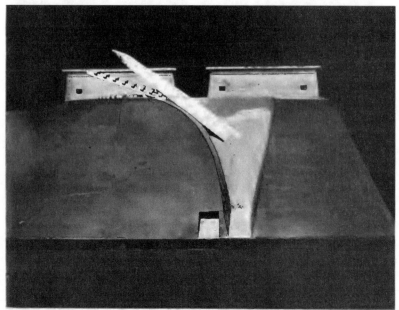

Engelbach's sectional model with the obelisk in the tunnel while sand is removed.

there are two rounded depressions near the centre. These may have been for inserting soft wooden blocks to act as shock-absorbers and to prevent the obelisk from tilting itself upright, prematurely, in its descent. The curious marks on the pedestal of the west obelisk of Luxor Temple may have fulfilled a similar purpose.

It is noteworthy, in the pedestals of the various obelisks, that their notches are not on the river side of the pedestals, even, as in the case of the obelisk which once stood before Pylon VII at Karnak, when the distance to the river was nearly 400 yards. To obtain sufficient height in the embankment, this obelisk had to be taken directly inland and brought back on an embankment which must have been constructed right over the Sacred Lake. Existing pylons prevented it from being brought to its pedestal parallel to the river, as was done in the case of the obelisks of Tuthmosis I and III on the axis of the temple at Karnak. This is another hint that the embankment theory is correct.

Engelbach says this about the thickness and extreme length of the largest obelisks, specifically the Unfinished Obelisk:

156

Before we can say that the funnel theory is a possibility, we have to make sure that the largest obelisk known will not break owing to its great weight when supported at or pivoting round its center of gravity or balancing-point. The non-technical reader will grasp this point better if he realizes that a model obelisk like that shown in the photographs [reproduced here], which can be supported anywhere, and even leaned upon as well, without breaking, will not behave in the same way if it is magnified some 200 times, although the proportions are identical and the material the same. The strain due to its own weight is proportional to the linear dimensions of the monument. I will not give here the extremely wearisome calculation for the strain set up, but it is given in full in *The Aswan Obelisk*, and shows that even it could be supported anywhere without straining the granite to more than two-thirds what it can possibly stand. This is a narrow enough margin, and to endure this strain the granite would have to be flawless. Although the mathematics of the Egyptians was totally incapable of determining such stresses, they knew very well that such a long obelisk, if not perfectly sound, would inevitably break during the erecting process, if not long before. One, called Dhutiy, mentions in his tomb inscription at Thebes that he erected two obelisks of 108 cubits in length, but unless the obelisks were much thicker than all known examples in proportion to their length, they would not have stood the strain of transport and erection.

He mentions the theory of the French Egyptologist Auguste Choisy, and what is pretty clear is that everyone has their own ideas on how the Egyptians might have raised an obelisk, something that was never described by the ancient Egyptians for unknown reasons. Perhaps they did not know how to raise an obelisk either. Says Engelbach on Choisy:

> One of the more surprising theories on the erecting process, which savors somewhat of Heath Robinson's mechanical studies, may be of interest. This is that put forward by Auguste Choisy in *L'Art de batir chez les Egyptiens*.
> According to him the obelisk was raised by a series of weighted horizontal levers acting along its length, earth

being banked under the obelisk at each heave, suitable supporting surfaces for the fulcra of the levers, in the form of masonry sides to the bank, being made, and heightened as the obelisk rose. [Choisy's illustration] taken from his book, makes this clear.

...He says: "Having arrived at a height a, let us pass, below it, cross-beams c and a pivot (tourillon). Now nothing prevents us from getting rid of the earth and constructing a glissiere, or slide, g. Having made the slide, let us replace the removed earth by sand; let us remove the supports c and take away the sand..."

... He does not tell us what the tourillon is to be made of to stand the enormous strain, nor does he give any details as to the nature of the slide which would allow the point of the sled to slide over it and not jam hard. His attachment of the obelisk on the sled and the recess in the latter for holding back the obelisk are quite unsupported by any evidence. He goes on to say that the obelisk was lowered down on to its

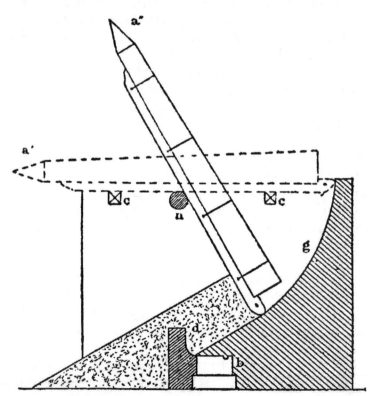

Engelbach's diagram of Choisy's proposed method for lowering an obelisk.

pedestal by puncturing filled sand-bags which had been packed between it and the pedestal when in "position a."

His explanation of the notch is that it was to take a sausage-shaped bag, which was to be punctured last, after having removed the debris of the others. The mechanics of the method seem to me to be quite unsound, and the crushing of the inner edges of the pedestal-notches and the position of Hatshepsowet's obelisk on its pedestal are not explainable by Choisy's theory.

Engelbach has several diagrams illustrating Choisy's theory, but it all seems a bit complicated and it seems that Engelbach prefers the funnel theory for erecting an obelisk. Indeed it seems the big "problem of the obelisks" is the mystery of how they were erected. Erecting an obelisk is very difficult without large cranes or some sort of levitation device. How did the Egyptians — and others — erect their massive obelisks? Why was the erection of obelisks never described?

Engelbach then gives us a description of a sacred barge and two great obelisks that were set before the barge:

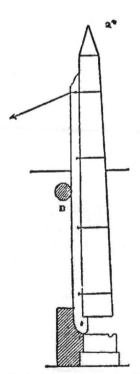

> We have the description of the furniture provided for the sacred barge of the god Amun in the time of Amenophis III. We are told (Breasted, *Ancient Records, II*, § 888): "It was made very wide and large; there is no instance of the like being done. It... is adorned with silver, wrought with gold throughout; the great shrine is of electrum, so that it fills the land with its brightness; its bows are as bright. They bear great crowns, and serpents twine along its two sides to protect them. Flagstaves, wrought with electrum are set up before it, with two great obelisks between them; it is beautiful everywhere."

Engelbach then says we know very

Choisy's proposed method for lowering an obelisk.

little about the polishing and engraving of obelisks:

> I had intended to devote a chapter to the polishing and engraving of obelisks after they were set up, but our knowledge of the engraving of the hard rocks is so vague that it can be summed up in a paragraph. The details of the processes, as given in the various works on the subject, are not clear to me—perhaps owing to my reprehensible habit of making experiments. The fundamental principles are, however, tolerably plain, and are summed up in Prof. Petrie's *Arts and Crafts in Ancient Egypt*. There is no doubt that the faces of the obelisks were dressed by the dolerite balls until they were as flat as possible, tests being made, as in engineering work today, by putting against them a portable flat plane smeared with red ochre and oil, or "ruddle" as the red lead and oil, now used for this purpose, is called. Prof. Petrie says that it was considered flat enough if the touches of red ochre from the plane were not separated by more than an inch, but I think he means this to refer to the sarcophagi and medium-sized monuments. In an obelisk the accuracy seems to have been far less. The basis of the polishing and the engraving was most certainly emery stone and powder. There are indications that granite was cut with tubular drills and sometimes sawn, but we are more than doubtful how the emery was used.

Engelbach then quotes page 72 of Petrie's book where he suggests that a gritty sand of emery powder was used with a copper saw:

> The difficult question is whether the material (emery) was used as loose powder, or was set in the metal tool as separate teeth. An actual example was found in the prehistoric Greek palace of Tiryns. The hard limestone there has been sawn, and I found a broken bit of the saw left in a cut. The copper blade had rusted away to green carbonate, and with it were some little blocks of emery about a sixteenth of an inch long, rectangular, and quite capable of being set, but far too large to act as a loose powder with a plain blade. On the Egyptian examples there are long grooves in the faces of the cuts of both saws and drills; and grooves may be made by

working a loose powder. But, further, the groove certainly seems to run spirally round a core, which would show that it was cut by a single point... The large hieroglyphs on hard stones were cut by copper blades fed with emery, and sawn along the outline by hand; the block between the cuts was broken out, and the floor of the sign was hammer-dressed and finally ground down with emery.

In the next chapter Engelbach relates some fascinating figures from ancient texts on the number of workers used to transport large granite statues or sarcophagi. Says Engelbach:

Some idea as to the number of men employed on the transport of stone can be obtained from the following three accounts of expeditions. King Menthuhotpe IV, of the XIth dynasty, sent an expedition to the Wady Hammammat to quarry stone for a large sarcophagus, and it is recorded that 10,000 men were sent out there. We are further told that it took 3,000 sailors from the Delta Provinces to remove the lid, which measured 13 feet 10 inches by 6 feet 5 inches by 3 feet 2 inches deep, from the quarry to the river. The "sailors " were probably a pressed gang of the amphibious inhabitants of the Delta lakes. The expedition seems to have been fortunate, as we are told that not a man perished, not a trooper was missing, not an ass died, and not a workman was enfeebled (Breasted, *Ancient Records*, I, § 448). In the reign of King Amenemhet III, of the XIIth dynasty, an official, also called Amenemhet, was sent to the same spot for 10 statues, each 8 feet 8 inches high. The personnel was made up as follows (Breasted, *Ancient Records*, I, § 710):

Necropolis soldiers   ...20
Sailors Quarrymen   ...30
Troops   ...2,000

Under Ramesses IV a large expedition was again sent to the Wady Hammammat for monumental stone. It numbered 8,362 persons, and consisted of:

High Priest of Amun,
Ramesses-nakht, Director of Works ...1
Civil and military officers of rank ...9

Subordinate officers	…362
Trained artificers and artists	…10
Quarrymen and stonecutters	…130
Gendarmes	…50
Slaves	…2,000
Infantry	…5,000
Men from Ayan	…800
Dead (excluded from total)	…900
Total:	8,362

It will be seen from these figures that huge numbers of men were sent far afield for monuments much smaller than the Aswan obelisk. It seems to have been the custom to use troops on this unpleasant kind of fatigue. It might be observed by the facetiously-minded person that the present-day unpopularity of all recruiting measures in Egypt is but an inherited race-instinct. As there was always a garrison at Aswan, large numbers of men would be available at very short notice. Another point in the above list is the relatively small proportion of actual quarrymen and stonemasons. Since the rock in the Wady Hammammat was basalt—and very hard—it is more than probable that the extraction of the monuments was done by pounding, and that the quarrymen and stonemasons were only needed to direct the unskilled laborers and to perform the skilled work, such as making the wedge-slots when necessary and to examine the quality of the rock. How much finishing was done out in the desert we have no means of knowing.

Engelbach then says this about a rather large, newly-finished obelisk mentioned in the Papyrus Anastasi I:

In a papyrus known as the Papyrus Anastasi I, which is a kind of collection of model letters for scribes to copy, one scribe called Hori writes to another called Amenemope hinting that he is not up to his job. He says (Gardiner, *Egyptian Hieratic Texts*, § XIII):
"An obelisk has been newly made… of 110 cubits (190 feet); its pedestal is 10 cubits (17.5 feet) square, and the block of its base makes 7 cubits in every direction; it goes in a slope (?) towards the summit (?) one cubit

one finger, its pyramidion is one cubit in height, its point measuring two fingers. Combine them so as to make them into a list, that thou mayest appoint every man needed to drag it... ."

Here the obelisk is extremely long, with a ridiculously short pyramidion, and the problem is an impossible one to solve for anyone who is not acquainted with the results of previous work in the quarry, and who is not familiar with the ground to be covered. The figures given are only sufficient to determine the weight of the obelisk. If such a problem was a typical one that scribes had to solve, the conclusion is that some kind of statistical record was kept in the archives of the various seats of learning to which the scribes had access. In other words, the experience of previous undertakings was at the disposal of the scribes.

About Greek and Roman texts concerning obelisks Engelbach says that we can't learn much but that a canal was dug during the later Ptolemaic Greek period to move an obelisk that was already fallen and lying on the ground. Says Engelbach:

Greek and Roman writers throw very little light on the transport and erection of large monuments except in giving dimensions of the blocks transported. Herodotus, in Book II, Chapter 175, tells us that King Amasis II brought a building of one stone from Elephantine which measured 34 feet 7 inches by 23 feet by 13 feet externally, and 30 feet 10 inches by 20 feet by 8 feet 4 inches internally, and that the 2,000 men appointed to convey it—who we are told were all pilots—took three whole years to perform their task.

Pliny, in his *Natural History, Book XXXVI*, Chapter 14, gives a slightly more valuable account of how King Ptolemy Philadelphus had an obelisk transported to Alexandria. He tells us that it was done by digging a canal from the Nile to the spot where the obelisk lay, passing below it, so that the obelisk was supported on either bank. Two large barges loaded with stones were unballasted below the obelisk which, rising, received its weight.

This may well have been true, but it was not the way in which the Egyptians transported them, for there is no trace of a canal near the Aswan quarries.

Indeed, digging canals to the site of the obelisk seems like a good way to go, considering their tremendous size and weight. But Engelbach points out that no canals were dug to the Aswan quarries and the obelisks had to be moved in some other way.

On the lack of records about quarrying, moving and erecting obelisks left to us by the ancient Egyptians themselves Engelbach says:

> The Egyptians, as it has already been remarked, have left us practically no information at all as to how they erected their obelisks. There is, however, a passage in the Anastasi Papyrus which refers to the erection of a colossus, and which is perhaps worth recording here, since it is fairly certain that the principle of the erection of the larger colossi was very similar to that of the erection of an obelisk. The text gives:

> "It is said to thee: *Empty the magazine that has been loaded with sand under the monument* of thy Lord, which has been brought from the Red Mountain. It makes 30 cubits stretched on the ground and 20 cubits in breadth... with 100 chambers (?) filled with sand from the riverbank. The ...of its chambers have a breadth of 44 (?) cubits and a height of 50 cubits, all of them ...in their... Thou art commanded to remove (overturn) it in six hours."

> Here, owing to errors in re-copying, and our slight knowledge of the technical terms mentioned, we are at a total loss as to the meaning of the second sentence.

> In the same papyrus (§ XIII) there is a reference to an embankment which may well have been intended for the erection of an obelisk, as the problem immediately following it is that dealing with the transport of an obelisk, which has already been quoted.

> The scribe Hori puts the problem thus:

> "There is a ramp to be made of 730 cubits (418 yards) with a breadth of 55 cubits (31.5 yards) consisting of 120 compartments (?) filled with reeds and beams having a height of 60 cubits (34.4 yards) at its summit. Its middle is 30 cubits (17.2 yards), its batter 15 cubits (8-6 yards), its base (?) 5 cubits (2.87 yards). The quantity

of bricks for it is asked of the commander of the army. Behold its measurements are before thee; each one of its compartments is 30 cubits long and 7 cubits broad...."

Here, as before, the words "compartment" and "base" are of very doubtful meaning, and it is difficult to arrive at any definite idea on the construction of the ramp apart from its overall measurements. However one tries to arrange compartments in the ramp, an impossible situation follows, so we are compelled to believe that there is some error in the figures due to re-copying. It is likely that the compartments refer to the internal division of the ramp which, as it were, is a brick box, filled with earth for economy; on the other hand, the word may mean the externally visible sections or towers always found in very large brick walls. For full notes on these walls, see Somers Clarke, *Journal of Egyptian Archeology*, Vol. VII, p. 77.

The only account of the erection of an obelisk by the Egyptians is that given by Pliny, which cannot fail to appeal to those who have had the fortune (?) to fall into the hands of an Egyptian dragoman. He must have livened up the visitors even in those days. Pliny was told that King "Rhamsesis," when an obelisk was being put up, feared that the machinery employed would not be strong enough, so he had his own son tied to the summit in order to make the workmen more careful. If this "Rhamsesis" was Ramesses II, the loss of a son would not have been vital, as he is known to have had over a hundred, to say nothing of several score daughters!

Engelbach then remarks on the strange and very tall obelisks of an architect named Dhutiy who claims, probably falsely, that he erected two great obelisks, taller than the Unfinished Obelisk. Says Engelbach:

Another obelisk-architect under this queen was one Dhutiy, whose tomb (No. 11 at Thebes) has been mutilated by Tuthmosis III. Among his many titles were Director of Works and Controller of the Double-houses of Silver and Gold. The great work by which he is known is the systematic recording of the treasures from the Punt expedition, and he appears—busily taking notes—in the reliefs in the temple of Der El-Bahari. As has been remarked, he was openly of the queen's party, and suffered in consequence. In addition to his recording work, he appears to have made gateways, shrines, thrones and small furniture for the temple of Karnak, and erected two great obelisks of 108 cubits (186 feet) high. We have no idea at all as to where these obelisks were placed; further, it seems that such a high obelisk could not withstand its own weight during its transport and erection (p. 76), unless it was vastly thicker proportionately than all others, so it has been suggested that the length given is the total length of the pair when placed butt to butt on the giant barge. It is more likely that the figure is an error in transcription from the cursive notes from which the tomb-inscriptions were copied.

The Unfinished Obelisk at Aswan is 137 feet in length and would have weighed approximately 1,168 tons. An obelisk that was 186 feet in length would weigh in excess of 1,500 tons and be the largest monolithic stone ever erected. As he says, he has no idea where this giant obelisk was allegedly erected, if ever. The incredible scene of a lost obelisk lying broken in the sand somewhere—larger than any known obelisk—is conjured by this mysterious statement. Was it true?

Engelbach also speaks of an engineer for Queen Hatshepsowet named Puimre and several missing obelisks:

> Puimre, whose name has already been mentioned in connection with Sennemut, although he had done certain pieces of work for Queen Hatshepsowet, managed to retain the favor of Tuthmosis III when he reigned alone. In his tomb (No. 39 at Thebes, lately restored) he states that he erected two obelisks for Tuthmosis III at Karnak. By a process of elimination, it is likely that they were those which stood before Pylon VII at Karnak. Judging from the base measurements of the eastern fragment, they must have stood between 94 and 115 feet high, that is, higher than the great obelisk of Hatshepsowet at Karnak, and only equaled by the Lateran obelisk at Rome.

So, clearly, a normal obelisk measured from 90 to 115 feet. Most are much smaller than that. The Unfinished Obelisk is already massively huge at 137 feet tall and an obelisk that was 180 feet tall would be a nearly impossible engineering feat. There is some doubt as to whether these obelisks were ever placed at Karnak as Puimre claims; it seems likely that if they were ever there, they would have been placed there before the temple was built. Many of the claims

Engelbach's photo of a statue of Sennemut, the architect of Hatshepsut's obelisk and holding the queen's daughter, Nephrue.

167

that are found in inscriptions and papyrus texts are suspect and Egyptologists are aware that many of the claims are obviously false. This seems to also include claims about obelisks.

Is it possible that obelisks are so old that they were already standing when temples were built around them? Is it possible that many of the obelisks were so old and without inscription that it was easy for a later pharaoh and his architects to claim them for their own? Engelbach concludes that the ancient Egyptians had a very sophisticated knowledge of engineering and the movement of large blocks of stone. That there is a serious lack of information on the creation and movement of obelisks seems to bother Engelbach but he forges ahead anyway.

In the final chapter Engelbach talks about the various removals and erections of obelisks in modern times, including those in Paris, London, New York and the Vatican. On the erection of the Vatican obelisk he tells us:

> The Vatican obelisk had been taken from Egypt in Roman times, and it was moved in A.D. 1586 by Domenico Fontana from the Circus of Nero at Rome to the Piazza di San Pietro, where it now stands, incongruously decorated—like most of the other obelisks in Italy—with a brazen cross. The removal was performed by order of Pope Sixtus V. The method used was the heroic one of lifting it bodily by systems of pulleys actuated by a large number of capstans. The pulleys were slung from a gigantic tower of wood, popularly known as "Fontana's Castle," which was made of compound wooden baulks over a yard square in section. The pulleys were attached to the obelisk at four points along its length, the inscriptions being protected by matting and planks. The obelisk was first raised sufficiently high, being wedged at the same time from below, to enable a "cradle," or platform on rollers, to be introduced underneath it. It was then lowered on to the cradle and pulled to its new site, first down an inclined plane and thence on level ground. The erecting was performed in exactly the reverse manner to the lowering. The whole story, as translated by Lebas in his *L'Obelisque de Louxor*, is distinctly diverting, and I cannot resist giving two extracts. He tells us (p. 178):
>
> "Public curiosity... attracted a large number of

strangers to Rome, and a bando of the Pope, published two days before, punished by death anybody who did not respect the barrier... On the 30th April, two hours before daylight, two masses were celebrated to implore the light of the Holy Spirit. Fontana, with all his staff, communicated... On the eve of the lowering he had been blessed by the Holy Father..."

Before the work began Fontana told his workmen: "The work we are about to undertake is consecrated to religion, the exaltation of the Holy Cross"; thereon everyone recited with Fontana a Pater and an Ave. The ceremony was made interesting for the spectators by the presence of some "familiars" of the Church, whose duty it was to administer summary punishment to anyone who misbehaved. Absolute silence for work-men and spectators was ordered, and the story is still told of a workman who disregarded the order at a critical moment, when the ropes had become slack and could be tightened no further. He cried: "Wet the ropes!"— which was done, and the situation saved. For his initiative he is said to have had an annuity granted to himself and his descendants by the Pope.

On the Paris obelisk Engelbach tells how this still-standing obelisk was lowered onto a wooden cradle and then dragged to the Nile:

Engelbach's photo of a model for erecting an obelisk.

169

The removal of the obelisk from Luxor Temple to the Place de la Concorde in Paris is perhaps the worst of these gross acts of vandalism, since the Luxor obelisks were the only pair still standing in their original position. It was done by an engineer called Lebas in 1836. The obelisk was lowered and raised by means of a huge compound sheers, consisting of five members, or struts, on each side of it. The power was supplied by systems of pulleys worked by capstans. The model ... makes this method clear as regards the appearance and position of the sheers, and the way in which the obelisk was slung from them, but only one capstan and system of pulleys is shown here. The obelisk was lowered on to a wooden cradle, on which it was dragged over a greased way, without rollers, to the Nile.

There a pontoon-raft, with its prow temporarily removed, was waiting to receive it. The raft was towed home, the prow again removed and the obelisk dragged to the Place de la Concorde on its cradle, being finally brought up a slope leading up to the surface of the high pedestal on which it was to be erected. Though the obelisk weighed but 227 tons, it took a pull of 94 tons from the capstans to move it up the gradual incline. The edge of the obelisk was made to rest over the pedestal-notch, in which it engaged as it rose towards the vertical. Lebas's book, which is now

Engelbach's photo of a model for erecting an obelisk.

very rare, is extremely interesting, giving many delightful sketches of some of the ludicrous situations met with in the course of the work, and of the cheery way in which the party overcame their difficulties, which ranged from an epidemic of plague to a shortage of wood.

About the New York obelisk Engelbach says that it was one of two lying in the sand near Alexandria and may have been moved there by the Romans two thousand years earlier in an effort to bring them to Rome. Though other obelisks were taken to Rome, these two were not. Where they originally came from no one seems to know. Says Engelbach:

> The New York obelisk originally formed a pair with the London obelisk in a temple at Heliopolis, near Cairo, and both had been moved in Roman times to Alexandria, close to the beach. The English took the one which was lying in the sand, leaving the Americans the other, which was standing on its pedestal. At an earlier stage of its history all four edges had been broken away, and four copper cramps—shaped like sea-crabs—had been put at the corners to support it more firmly. In modern times only two of the crabs remained, the others having been stolen and blocks of stone put in their stead.
>
> The method of lowering the obelisk was ingenious in the extreme. The obelisk was first fitted with a pair of huge steel trunnions (similar to those seen on a toy cannon by means of which it can pivot around its centre). The trunnions were left loose until two steel towers had been constructed on either side of the obelisk, as shown in the model, to act as a support for them. A strong steel plate was passed under the butt of the obelisk and attached by a series of stout steel bars or" tension-rods," which could be shortened by screwing. Whether there was originally a space below the center of the butt, or whether the obelisk was raised by jacks or rams placed under the four rounded-off corners, I am uncertain.
>
> The tension-rods were shortened by screwing, and the obelisk thus pulled clear off its pedestal, being supported by, and sliding through, the trunnion. The trunnion, which was arranged to be at the balancing-point of the obelisk when it was sufficiently high, was next bolted tight and the obelisk

itself braced by long rods, passing, as shown in the model, over a stiff support at its centre. From this position it was intended to let the point of the obelisk come slowly round until it rested on a crib of wooden baulks.

What actually happened was that, owing to a miscalculation of the balancing-point, the tip crashed down, breaking the holding-back ropes. It splintered about

Thutmosis III presenting obelisks and flagstaffs from Palestine to Amun at Karnak.

172

three courses of baulks and escaped breaking by a miracle. Another crib of baulks was next built below the butt. The next step was to remove the towers and the trunnions; this was done by taking the weight of the obelisk off them by raising the point by oil-rams placed within the wooden crib. For those unacquainted with rams, it may be explained that they are appliances by which a great lifting force can be obtained for a short distance by means of oil compressed into them by a pump. A" jack," which enables one man to lift up the back of a heavy motor, has a similar function. In the model shown, the jack is actuated by hand through a bowden wire.

[The next figure] shows the weight of the obelisk being taken by the ram, so that the towers and trunnions can be removed. This being done, the rams are released and the obelisk comes down on to the crib. The rams are then used from each crib in turn, lifting the tip or butt so that a course of baulks can be removed and the obelisk gently lowered on to the course below. [The next figure] shows the obelisk when it has nearly arrived at the ground. It had originally been intended to convey the obelisk through the streets of Alexandria to the harbor, but the inhabitants, especially the European community, who had opposed the removal strenuously, influenced the Municipal Council to forbid this. A special wooden slide had therefore to be constructed so that the obelisk, which was to be put in a wooden caisson,

Engelbach's photo of a different model for erecting an obelisk. **173**

could be pulled down it to the sea, and floated round to the harbor instead. At the harbor it was introduced into a steamship called the *Dessoug*, by opening a port in her bows. The journey to America was comparatively uneventful, and between the harbor and Central Park it did the longer journeys by rail and the shorter journeys rolling on cannon-balls running in U-shaped "channel-irons"; i.e., cannon-balls were used as ball-bearings! At Central Park the erection was performed, with elaborate ceremonial, under the auspices of the Freemasons. The method of erection was exactly the reverse of that used for the lowering, and it was carried out without a hitch on January 22, 1881, or just about 2 and 1/2 years after the London obelisk was set up. The work done was under the direction of Lt. Commander H. H. Gorringe, U.S. Navy.

Finally, Engelbach tells us about the sister obelisk lying on the

Ramses II showing obelisks, flagstaffs and colossi at Luxor Temple.

174

beach at the Mediterranean port of Alexandria which was taken to England and became the London obelisk. This obelisk was lost at sea but then recovered:

> The London obelisk had only to be transported and erected, since it was already lying unbroken in the sand at Alexandria. The principle of the erecting process was the same as that used for the New York obelisk, except that, instead of the trunnions, steel shoulders with "knife-edge" bearing surfaces were used. These engaged in a huge wooden scaffolding instead of on the two steel towers. For transporting it by water it was enclosed in a steel shell, fitted, like a ship, with deck and mast. It even had watertight compartments. The "ship" was named the *Cleopatra*, and she set out from Egypt on the 21st of September, 1877. She steered very badly, and in a gale near Cape St. Vincent the steamship *Olga*, which was towing her home, had to cut the "august barge" adrift. Six sailors, who tried to reach the *Cleopatra* to secure her ballast, perished in the heavy sea. The *Olga* then lost the *Cleopatra*, and, imagining she had foundered, she steamed home. The *Cleopatra*, however, had not foundered at all, and was salved by a ship called the *Fitzmaurice*, who towed her into Ferrol. A claim for £5,000 salvage was reduced by the Admiralty Courts to £2,000.
>
> Having arrived in the Thames on January 20th, 1878, the obelisk was brought right up beside the site on the Thames Embankment where it now stands, being grounded at high tide. After the shell had been cut away, the lifting on to the Embankment was done almost entirely by hydraulic jacks. At its erection, which took place in September, 1878, an extraordinary collection of objects was put in the base of this obelisk, which ranged from sets of coinage, newspapers and standard works, to a Mappin's shilling razor, an Alexandra feeding-bottle, a case of cigars and photographs of a dozen pretty Englishwomen for the benefit of posterity! What would the feelings of Tuthmosis III have been when he ordered these obelisks for the god Re, had he known that one would be taken to a land of whose existence he never dreamed, and that the other would fall into the hands of what was then a savage people, and, after undergoing such vicissitudes as shipwreck and injuries from a German air-

bomb, would still be standing, though thousands of miles away, after a lapse of nearly 3,500 years?

Engelbach ends his book with several appendices including a list of the various Egyptian kings that are mentioned in the book, which is quite a few. Not all of them are said to have erected obelisks, and in fact, that list is probably pretty short.

For over 50 years Engelbach's book was virtually the only book written in English about obelisks. Then Peter Tompkin's book *The Magic of Obelisks* was published in 1981, and it briefly focused worldwide attention on obelisks.[2] What are obelisks really for? Are they just symbols of the sun and monuments to phallic-inclined engineers?

The monolithic granite obelisks of the past were not a symbol of the sun, they were a symbol of the awesome ability of the kings and their engineers who could quarry, move and erect such colossal monoliths weighing many hundreds of tons. The people who would marvel at these superb pieces of engineering would be other engineers and architects—a noble profession, even in ancient Egyptian times.

These obelisks are symbols of the superior intellect and engineering ability of their builders, who deserve tremendous credit for their exploits. But is it all just about bragging rights about who can put up the largest obelisk and do something that other cultures can't do? Or is there a greater purpose that these towers of stone had as their builders intended? Was there a reason other than attempting—and creating—something that is terribly difficult in order to impress people? Were these obelisks part of a worldwide energy system in the past? Were they some sort of transmitting tower or antenna? Let us look at obelisks around the world, beyond their supposed homeland in Egypt.

Chapter 5

Obelisks in Ethiopia

It is therefore a truism, almost a tautology,
to say that all magic is necessarily false and barren;
for were it ever to become true and fruitful,
it would no longer be magic but science.
—Sir James Frazer, *The Golden Bough*

While Egypt is famous for its obelisks, the city of Axum in Ethiopia is also noted for its huge obelisks. In fact, the largest quarried obelisk that was once standing (the Unfinished Obelisk at Aswan is bigger) can be found in Ethiopia, now smashed into gigantic pieces. The city of Axum is associated with the Queen of Sheba and King Solomon and there is a certain amount of debate as to whether the obelisks of Axum were there during the period of Solomon and Sheba. Ethiopians believe that the obelisks are many thousands of years old and were present during the time of the Queen of Sheba, circa 1000 BC. They even believe that the power of the Ark of the Covenant was used to erect the towering monoliths.

The Ethiopian holy book, the *Kebra Nagast*, gives us information on the Ark of the Covenant, a flying wagon, and other magical things, but does not mention the obelisks of Axum. Egyptian texts rarely mention obelisks either, even though they have been standing in Egypt for at least four thousand years, if not more. Commonplace monuments, including large buildings and obelisks are often not mentioned in texts. There is little to say about them anyway since most people do not know their origin or purpose, even though they may see them nearly every day.

The Building of the Temple

The *Kebra Nagast* tells us that Solomon gave Sheba an airship, and later his son Menelik had one as well. While the Bible does not tell us about Solomon's airship, it does tell us that he and his kingdom became fabulously rich by sending a fleet of ships to a mysterious land called Ophir to bring back over 20 tons of gold, spices and other goods.

We learn about King Solomon in the Old Testament books 1 Kings and 2 Chronicles. 1 Kings gives the general story of Solomon and of how Israel becomes powerful during his reign. It also tells of Solomon's death and the division of Israel into two states, Israel and Judah. 2 Chronicles continues to recount the life of Solomon and the building of the temple.

In the first chapters of 1 Kings, Solomon becomes the king of Israel. In 1 Kings 3 he makes an alliance with the Egyptian pharaoh, marrying his daughter, and asks for wisdom from god in a dream.

In 1 Kings 5 we are told that the Phoenician King Hiram of Tyre (in what is today Lebanon) sends envoys to Solomon when he hears he has become king. Solomon wrote a letter to be taken to Hiram by the ambassadors stating his intention to build "a temple for the name of the Lord" now that there was relative peace in the region and he could turn his attention to this task. He asks that the cedars of Lebanon be cut for him, and sets up a rotation system for Israelite

An old print of the building of Solomon's temple.

conscripted laborers to spend time in Tyre working alongside Hiram's expert wood- and stoneworkers. Says 1 Kings 5:17-18:

> At the king's command they removed from the quarry large blocks of high-grade stone to provide a foundation of dressed stone for the temple. The craftsmen of Solomon and Hiram and workers from Byblos cut and prepared the timber and stone for the building of the temple.

In 1 Kings 6 we are told more about Solomon building the great temple which was to house the Ark and impress anyone who entered it with the greatness of Yahweh. The temple was furnished with much gold and adornments and took seven years to build. In 1 Kings 6:7 we are told that no hammers or iron tools were used at the temple site:

> In building the temple, only blocks dressed at the quarry were used, and no hammer, chisel or any other iron tool was heard at the temple site while it was being built.

Because of this curious statement, a legend rose up that King Solomon possessed a magic substance—or "worm"—the size of a barleycorn, that could cut stone. This magical item was called the Shamir and it could be used to cut any kind of metal or stone and was essentially harder than any other substance.

It is difficult to find a lot of information on the Shamir but let us look at the Wikipedia entry under "Solomon's Shamir":

> In the Gemara [part of the Talmud], the Shamir is a worm or a substance that had the power to cut through or disintegrate stone, iron and diamond. King Solomon is said to have used it in the building of the First Temple in Jerusalem in the place of cutting tools. For the building of the Temple, which promoted peace, it was inappropriate to use tools that could also cause war and bloodshed.
>
> Referenced throughout the Talmud and the Midrashim, the Shamir was reputed to have existed in the time of Moses. Moses reputedly used the Shamir to engrave the Hoshen (Priestly breastplate) stones that were inserted into the breastplate. King Solomon, aware of the existence of the Shamir, but unaware of its location, commissioned a search

179

that turned up a "grain of Shamir the size of a barley-corn."

Solomon's artisans reputedly used the Shamir in the construction of Solomon's Temple. The material to be worked, whether stone, wood or metal, was affected by being "shown to the Shamir." Following this line of logic (anything that can be 'shown' something must have eyes to see), early Rabbinical scholars described the Shamir almost as a living being. Other early sources, however, describe it as a green stone. For storage, the Shamir was meant to have been always wrapped in wool and stored in a container made of lead; any other vessel would burst and disintegrate under the Shamir's gaze. The Shamir was said to have been either lost or had lost its potency (along with the "dripping of the honeycomb") by the time of the destruction of the First Temple at the hands of Nebuchadnezzar in 586 B.C.

According to the deutero-canonical Asmodeus legend, the Shamir was given to Solomon as a gift from Asmodeus, the king of demons. Another version of the story holds that a captured Asmodeus told Solomon the Shamir was entrusted to the care of a woodcock. Solomon then sends his trusted aide Benaiah on a quest to retrieve it.

The Shamir worm was also used by King Solomon to engrave gemstones. Apparently he also used the blood of the Shamir worm to make carved jewels with a mystical seal or design. According to an interview with Dr. George Frederick Kunz, an expert in gemstone and jewelry lore, this led to the belief that gemstones so engraved would have magical virtues, and they often also ended up with their own powers or guardian angel associated with either the gem, or the specifically engraved gemstones.

So what was the Shamir? It is tempting to think it was some sort of laser technology, with the "eye" of the Shamir being the focused beam of light emitted from the laser, which could easily burn through stone, metal or wood. The first modern laser, developed in 1960, used a synthetic ruby crystal to beam red light, and the Shamir is associated with gems. Why did the Shamir have to be kept in a lead box? Lead is usually used as a protective layer against excess radiation. "Laser" stands for Light Amplification by the Stimulated Emission of Radiation, and radiation injury is possible even from moderately powered beams. But having a laser device

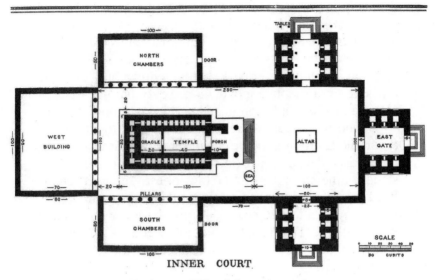

The layout of King Solomon's temple.

the size of a barleycorn seems far-fetched even by today's standards of miniaturization.

The general thought is that it is a diamond, the hardest substance in the mineral world. In fact, the Shamir may have been some sort of diamond drill or saw that would cut other gems and saw and drill through such hard stones as granite or even basalt. To be "shown to the shamir" may have been akin to having an object placed on a diamond lathe or saw. Today it is necessary to use diamond saws and drills to cut modern granite blocks at quarries and we might imagine that ancient stonecutters—some of them at least—had similar tools.

Again, was some highly developed civilization, or several of them, producing diamond saws, diamond drills, iron tools, electrical devices and even airships? Countries like ancient Egypt, Babylonia, the Hittite Empire, India and China were all capable, in my mind, of producing such things in ancient times. Other countries, such as ancient Israel, Greece or Ethiopia had to import these items and they were therefore "magical" objects and only owned by kings and important princes.

A fascinating website called BibleSearchers.com has some interesting information on the Shamir, which was most likely composed of small pieces of diamond, sapphire and/or ruby, probably from India:

In the Mishnah Avon 5:6, the Shamir was created on

181

the sixth day of creation and was given to the hoopoe-bird (woodcock) who kept it in her custody throughout the ages in the Garden of Eden. This marvelous bird would on occasion take this worm and carry it across the earth, carrying it tightly in her beak, letting it down only to create a fissure on a desolate mountain peak so that the seeds of plants and trees could sprout and provide her food.

When the Israelites were camped near Mount Horeb/ Sinai, the Lord brought the Shamir and gave it to Bezaleel to engrave the names of the twelve tribes on the twelve stones of the breastplate of the high priest, Aaron. Then the Lord gave it back to the custody of the hoopoe-bird. Here she kept it in a leaden box, with fresh barley, wrapped in a woolen cloth. That is until Solomon needed it to build the Temple of the Lord in Jerusalem. Since that day, the Shamir has been lost.

As with all good rabbinic Talmudic debates, there was always a dissent. Judah R. Nehemiah claimed that the stones were quarried and then brought to the temple in a finished condition for the building of the temple. It appears that Rabbi Nehemiah's argument carried the debate as most scholars today believe this also to be true.

Of course, most Talmudic arguments were debated during the Roman imperial rule. In Latin, the Shamir was known as *smirks corundum*, the substance of sapphires and rubies and the hardest known gem next to the diamond. The substance of legends has a kernel of truth and now we know the 'rest of the story.'

These Bible scholars think that there was more than one Shamir and they were none other than power tools, specifically diamond drills and such. Says BibleSearchers:

> The whole passage in the Book of Kings suggests that there was a certain dignity and quietness, a decorum that was to be maintained as the contractors and the artisans could feel the presence of the Lord in the House that they were constructing. There was to be a spirituality of the Presence. They were not to hear the pounding or banging of instruments of iron. Was the literalness of the message also to have a spiritual connection? Yet within the House

of the Lord, there [were] no injunctions of the Lord against the humming of diamond drill bits as they cut, polished and finished off the massive limestone walls, or trimmed the edges of the cedars of Lebanon, or engraved the wood on the porticoes, or drill holes into the limestone to set beams and stabilize pillars, to embellish the trimming on the ceilings of the Holy Place, place engraved images on the Molten Sea or on the large doors that entered into the temple proper. Is it not time to consider that Solomon with all his wisdom and wealth also had access to technologies that we think are modern, only to someday know that the ancients were using them too?

...Maybe within the hoard of the treasures of Solomon's temple, we will find evidence of the technological sophistication, such as diamond drills, diamond and corundum bit saws that scholars have long felt did not exist in the 11th century BCE.

An old print of the obelisks of Axum.

The actress Betty Blythe as the Queen of Sheba in a 1921 film.

BibleSearchers give a date for King Solomon between 1001 and 1100 BC, though others give the date at around 970 to 931 BC. No matter what the exact date for the building of Solomon's Temple, it was apparently a massive construction job, with some of the building consisting of huge ashlar stones, but these were likely already in place from a much earlier date.

Nearly all of this construction was done by Phoenicians from Tyre, Byblos and other Phoenician cities in the eastern Mediterranean. Indeed, it was Solomon's good fortune to have a Phoenician king who loved him and his countrymen and would go to great lengths to help Solomon build his nation into one of great prosperity.

184

The Queen of Sheba and Axum

The ships to Ophir are mentioned amid a description of the visit of the Queen of Sheba to Solomon's court. Although the comments about the navy seem to be an aside interrupting the narrative, we get the idea that the ships may have been stopping at a port on the Red Sea that was in her kingdom, and she heard the tales of King Solomon from the sailors. In 1 Kings 10:1-15 we are told:

> And when the queen of Sheba heard of the fame of Solomon concerning the name of the Lord, she came to test him with hard questions. And she came to Jerusalem with a very great train, with camels that bore spices and very much gold and precious stones; and when she had come to Solomon, she communed with him about all that was in her heart. And Solomon told her all her questions; there was not any thing hidden from the king which he told her not.

> And when the queen of Sheba had seen all Solomon's wisdom, and the house that he had built, and the meat of his table, and the sitting of his servants, and the attendance of his ministers, and their apparel, and his cupbearers, and his ascent by which he went up unto the house of the Lord, there was no more spirit in her.

> And she said to the king, "It was a true report that I heard in mine own land of thy acts and of thy wisdom. However I believed not the words until I came and mine eyes had seen it; and behold, the half was not told me. Thy wisdom and prosperity exceedeth the fame which I heard. Happy are thy men, happy are these thy servants, who stand continually before thee and who hear thy wisdom. Blessed be the Lord thy God, who delighted in thee, to set thee on the throne of Israel! Because the Lord loved Israel for ever, therefore made He thee king to do judgment and justice."

> And she gave the king a hundred and twenty talents of gold, and of spices a very great store, and precious stones; there came no more such abundance of spices as these which the queen of Sheba gave to King Solomon.

> And the navy also of Hiram, that brought gold from Ophir, brought in from Ophir a great plenty of almug trees and precious stones. And the king made of the almug trees pillars for the house of the Lord and for the king's house, harps also and psalteries for singers. There came no more

such almug trees, nor were seen unto this day.

And King Solomon gave unto the queen of Sheba all her desire, whatsoever she asked, besides that which Solomon gave her of his royal bounty. So she turned and went to her own country, she and her servants.

Now the weight of gold that came to Solomon in one year was six hundred threescore and six talents of gold, besides what he had from the merchants, and from the traffic of the spice merchants, and from all the kings of Arabia, and from the governors of the country.

Was the Queen of Sheba traveling from Axum? Where was Ophir? Where was all this gold coming from, not only Solomon's, but Sheba's?

The Land of Saba

It is quite apparent that the Queen of Sheba ruled over a rich country heavily involved in trade. In fact, the *Kebra Nagast* says that it was from a wealthy merchant that Sheba learned of Solomon.

For various reasons, many modern historians are doubtful that the Queen of Sheba existed, and if she did, they doubt that she was from Axum. I think they are wrong on both counts, and certainly millions of Ethiopians believe that she was a queen from Axum just as millions of Hindus in India believe that Krishna and Rama were real people.

The Queen of Sheba's visit to King Solomon was something of a trade trip, as the Queen brought with her some hundred and twenty talents of gold plus a multitude of spices and precious stones. As the Bible records, "there came no more such abundance of spices as these which the Queen of Sheba gave to King Solomon." The Queen also wished to test the wisdom of Solomon, and see for herself his magnificent court. As she leaves, Solomon gives her anything she wants, in addition to his royal bounty. "So she turned and went to her own country, she and her servants." (1 Kings)

So, the 120 talents of gold were largely for trade, with the Queen taking the "royal bounty" in return. Yet, scholars to this day are divided as to who the Queen of Sheba was and where she came from. She is not given a name in the Bible, just the title, Queen of Sheba, or Saba.

The Sabeans are mentioned in the Quran twice. The Quran mentions the Queen of Saba and names her Bilqis in the 27th chapter.

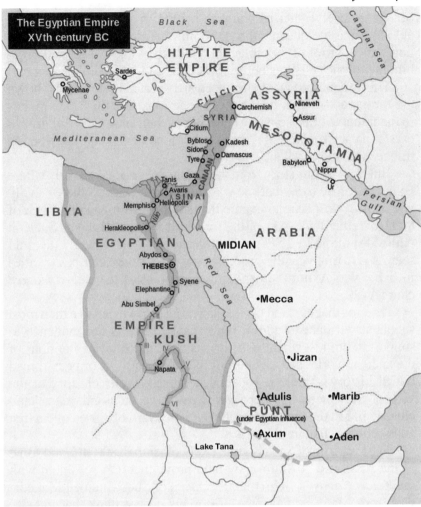

The Egyptian Empire
XVth century BC

In this chapter Suleiman (Solomon) gets reports from the hoopoe bird about the kingdom of Saba. It is ruled by a queen whose people worship the sun and planets instead of God. Suleiman (Solomon) sends a letter inviting her to submit fully to the One God, Allah, Lord of the Worlds.

The Queen of Saba is unsure how to respond and asks counsel of her advisors. They reply that the country is "of great toughness" in a reference to their willingness to go to war if necessary. She replies that she fears if they were to lose, Suleiman might behave as most other kings would: "entering a country, despoiling it and making the most honorable of its people its lowest."

She decides to meet with Suleiman and sends him a letter.

Suleiman receives her response to meet him and asks if anyone is able to bring the queen's throne to him before she arrives. A djinn under the control of Suleiman then tells the king that he will bring him the throne of Bilqis before Suleiman can rise from his seat.

This is done and when the queen arrives at his court she is shown the throne and asked: does your throne look like this? She replies to them that the throne was as her own. When she enters Solomon's crystal palace she accepts Abrahamic monotheism and the worship of one God alone, Allah.

But this curious story from the Quran does not say exactly where Bilqis is from, except that she was the queen of Saba. Where exactly was Saba? Most scholars agree that southwestern Arabia (much of today's Yemen) was part of the kingdom of Saba. But was Saba on both sides of the Red Sea, in Eritrea and Ethiopia as well? It would seem so. If this was the case, there would have been many cities in Saba; was Axum the capital at one point as the *Kebra Nagast* claims?

It seems that the land of Sheba was far more extensive than most scholars will dare to admit, not because there is not evidence to support it, but just because historians have preconceived notions of the ancient world, and this part of the world has sort of been forced out of history—Ethiopia especially. Indeed, the brief story of the visit of the Queen of Sheba gives us, three thousand years later, clues to life in the South of Arabia and the Horn of Africa in the first and second millennia BC.

The kingdom of Saba is quite likely older than 800 BC, and some of the structures at Axum may help prove that. One problem with the *Kebra Nagast*'s claim that the Queen of Sheba, named Makeda, was from Axum is that many historians do not think that the city existed at the time. I think those historians are wrong on this, and we will discuss the antiquity of Axum shortly. The stories of the *Kebra Nagast* predate the Quran, but it would seem that Mohammed was unaware of the book. However, there is some question as to when the stories were gathered together in one volume.

Though not much is known about ancient Saba, I contend that it was a huge country, with wealthy port cities on both sides of the Red Sea; some of the world's most impressive megaliths exist at Axum. There is good evidence that these megaliths are over three thousand years old, as we shall see.

That the queen's caravan was loaded with spices is only natural. Not only would the queen's realm have included the Dhofar region of the Hadramut, a major source of frankincense in the world, but it controlled the thriving trade with India and the spice islands of Indonesia. Saba controlled nearly all the production of frankincense, which is grown in southern Arabia, Ethiopia and northern Somalia. Today northern Somalia is the largest producer of frankincense, and the Vatican is the main buyer of the crop.

The lost city of Ubar was rediscovered in the early 1990s in Oman using satellite photography, and it is believed to have been an eastern center of the frankincense trade along the "Incense Road." Ubar and Oman are on the eastern edge of the frankincense production zone.

As the German scholar Joachim Leithauser once pointed out, if it were not for the incredible demand for spices and aromatics, almost all of which came from Southeast Asia or India, the world would never have been explored. Nutmeg, cloves, cinnamon, cardamom and other spices were literally worth their weight in gold. Frankincense, however, came from southern Arabia and the Horn of Africa.

Trade in spices had been going on for thousands of years before King Solomon's time and it seems only natural that the empire of Sheba was in control of a portion of that trade during the queen's time. The Babylonians and Persians had their own trading vessels that made the easy voyage, but it was Saba that controlled the trade in the Red Sea. Ports such as Adulis and Aden controlled traffic to the trading cities in Egypt, Israel and the eastern Mediterranean. Some of the major cities in Saba were Axum, Wiqro and Adulis in Ethiopia and Najran, Zafar, Marib, Aden, Timna and Qana in southwest Arabia.

The ancient kingdom of Saba—the land of Sheba—must have existed circa 1200 BC and even earlier. There seems little doubt that during the early dynasties of Egypt there was trade and contact going on with Ehiopia and Arabia, including maritime traffic. Saba would have included most of Southern Arabia and northern Ethiopia during the time of King Solomon. The question is whether Axum could have been the capital. Most historians say that the capital of Saba was at Marib, a desert oasis to the east of Sana'a in present-day Yemen. However, Marib was more likely a later capital of Saba when the country began to fall apart around the third century BC, some six hundred years after Solomon, Sheba—and their son Menelik—were exchanging gifts and flying over oceans.

The great dam at Marib is thought to have been built about 800

BC, before it became the capital of Saba; The area was conquered by the Himyarites around 100 AD. Marib seems like an unusual site for the capital of a united Saba, especially since some of its most important cities were ports on the Red Sea. It would seem that Saba was centered on Axum in its early days, and Marib became the dominant city later. The *Kebra Nagast* basically tells us this but historians have continually doubted the antiquity of Axum.

Thoth and the Baboons of Saba

Living in both Ethiopia and Yemen is the hamadryas baboon (dog-faced baboon), the northernmost of all the African baboons. This baboon is native to the Horn of Africa and the southwestern tip of the Arabian Peninsula, essentially the area of ancient Saba. The hamadryas baboon is smaller than other baboons. They live in semi-desert areas and require cliffs for sleeping so they can avoid predators, which are leopards or hyenas. The hamadryas baboon is omnivorous and is adapted to its relatively dry habitat. How did this baboon get on both sides of the Red Sea? Presumably via a land bridge, but they were sacred in Egypt and even kept as pets. Many tomb scenes show the animal led on a leash, or playing with the children of the household. It is believed that some baboons were trained by their owners to pick figs from the trees for them.

The hamadryas baboon appears in various roles in ancient Egyptian religion, and so is sometimes called the "sacred baboon." Baboons are not native to Egypt, however, and were imported from Ethiopia or Yemen. Hamadryas baboons were even trained as temple guardians or something similar. This was essentially the plot of Michael Crichton's book *Congo* (1980) which was made into a film in 1995, in which generations of trained gorillas have been guarding an Egyptian diamond mine that also contains temples.

The Egyptian god Thoth was often depicted in the form of a hamadryas baboon, sometimes carrying a crescent moon on his head. Thoth was also represented as an ibis-headed figure. Hapi, one of the Four Sons of Horus that guarded the organs of the deceased in ancient Egyptian religion, is also represented as having the head of a dog-faced baboon. Hapi protected the lungs and therefore a baboon's head is the lid of the canopic jar that held the lungs.

Thoth played many prominent roles in Egyptian mythology, including maintaining the universe, arbitrating godly disputes, and being one of the two deities (the other being Ma'at) who stood on either side of Ra's boat. Thoth was also associated with the arts of

190

Canopic jars used in Egyptian tombs.

magic, the development of science, and the system of writing.

As Thoth's sacred animal, the baboon was often shown directing scribes in their task. Baboons were said to carry out Thoth's duties as the god of measurement. They are sometimes seen with the scales that weighed the heart of the deceased in the judgment of the dead and are depicted at the spout of water clocks.

So we see that Egypt had close relations with not just Nubia to its south, but also to Ethiopia. Important rivers feeding the Nile begin in Ethiopia including the Blue Nile which originates at Lake Tana. When we look at Axum its similarities to Egyptian cities like Aswan or Luxor are astonishing. Like these cities, Axum has megalithic buildings and obelisks which are thousands of years old. Luxor Temple with its obelisks is dated to approximately 1400 BC. Could Axum be from this same time period?

The Antiquity of Axum

Among the many things in the *Kebra Nagast* that modern historians tend to discount is the antiquity of Axum and its monuments. The book describes events that supposedly took place circa 950 BC or earlier. Yet, the book starts at the Council of Nicea and is clearly a collection of works compiled around 400 AD or later. The accounts of flying wagons and the removal of the Ark of the Covenant from Jerusalem are all considered to be fictitious fantasy by modern scholars. Also, Axum could not have been the city of the Queen of Sheba as historians say that it was probably not in existence in 950 BC.

But, is it possible that the *Kebra Nagast* tells a story that is largely true? Certainly millions of Ethiopians believe the account

191

An early painting of Axum by Henry Salt, 1809.

to be a true version of what happened thousands of years ago. Is Ethiopia as important as the *Kebra Nagast* claims it is? Did they have some sort of flying vehicles at this time? Was the Ark of the Covenant—or a replica—removed from Jerusalem and taken to Ethiopia?

Graham Hancock, in his 1992 book *The Sign and the Seal*,[4] makes no mention of the flying wagons, but does discuss the Ark of the Covenant flying.

Hancock thinks that the Ark of the Covenant may have been taken from Jerusalem and then through Egypt, as described in the *Kebra Nagast,* and then brought to Lake Tana, the source of the Blue Nile. However, this did not happen during the time of Menelik but in the time just after the reign of King Uzziah (c.781-740 BC). Hancock thinks that Uzziah might have been standing in front of the Ark of the Covenant circa 730 BC when he was struck with leprosy as mentioned in 2 Chronicles 26: 16-21:

> But after Uzziah became powerful, his pride led to his downfall. He was unfaithful to the Lord his God, and entered the temple of the Lord to burn incense on the altar of incense. Azariah the priest with eighty other courageous priests of the Lord followed him in. They confronted King Uzziah and said, "It is not right for you, Uzziah, to burn incense to the Lord. That is for the priests, the descendants

192

There were two obelisks in front of the pylons at Luxor Temple, but today there is only one.

An aerial photo of the remarkable Er Grah obelisk in Brittany, mysteriously shattered in four pieces.

Above: An aerial photo of the Unfinished Obelisk. Below: Engineer Christopher Dunn examining the trench of the unfinished obelisk of Aswan.

Above: A photo of the Unfinished Obelisk.

Some of the standing obelisks at the central area of Axum.

One of the tall obelisks still standing at Axum.

Keystone cuts for metal clamps can be seen on ancient structures at Axum, just like in Egypt;

Portions of the broken Great Obelisk at Axum landed on this gigantic structure called the Tomb of Nefas Mawcha.

Portions of the broken Great Obelisk at Axum, once weighing 520 tons.

The Roman obelisk at Arles, France—one of a number of obelisks erected during Roman times. The Romans also imported Egyptian obelisks, but this one is of Roman origin.

Left: A natural quartz crystal from Brazil in the shape of an obelisk. In many ways granite obelisks are large versions of a quartz crystal.

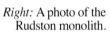

Right: A photo of the Rudston monolith.

of Aaron, who have been consecrated to burn incense. Leave the sanctuary, for you have been unfaithful; and you will not be honored by the Lord God."

Uzziah, who had a censer in his hand ready to burn incense, became angry. While he was raging at the priests in their presence before the incense altar in the Lord's temple, leprosy broke out on his forehead. When Azariah the chief priest and all the other priests looked at him, they saw that he had leprosy on his forehead, so they hurried him out. Indeed, he himself was eager to leave, because the Lord had afflicted him.

King Uzziah had leprosy until the day he died. He lived in a separate house —leprous, and banned from the temple of the Lord. Jotham his son had charge of the palace and governed the people of the land.

Hancock thinks that sometime shortly after the reign of Uzziah the Ark of the Covenant was taken to a Jewish community on Elephantine Island in southern Egypt and then to Lake Tana. It was kept for some time in a temple on one of the many islands in this large lake in the mountains of western Ethiopia; there are many ancient churches and temples on these islands to this day. Hancock thinks that the Ark would have arrived here sometime around 640 BC and eventually it was moved to Axum, where the Ethiopian Coptic Church says it resides today.

It is reasonable to think that the sacred ark of the Ethiopians

An early photo of the monuments at Axum, 1936.

did reside at Lake Tana (Hancock says specifically the island of Tana Kirkos), and now resides at the Church of Our Lady Mary of Zion in Axum. Perhaps the Ark was first brought from Jerusalem to Axum by Menelik and then later, during an invasion, it was taken to the island of Tana Kirkos. Later it was returned to Axum. But, is it the genuine Ark of the Covenant from Jerusalem or a copy? If it is genuine, why should we not believe the ancient book the *Kebra Nagast*? Just as we should believe certain accounts of the books of Exodus, Numbers and Kings, I think that we must believe that at least some of the things in the *Kebra Nagast* are true. Ethiopians would certainly want to think that!

The problem is that mainstream historians don't think that Axum was a capital in 950 BC, and prefer to date the Axumite kingdom to beginning around 400 BC. Mainstream historians think that the earliest monumental buildings in Ethiopia are those at Yeha, to the east of Axum, which are dated to circa 700 BC. So the question is whether Axum is older than Yeha as the *Kebra Nagast* claims.

First, let us look at Axum, its amazing monumental works and the Axumite dynasty. Axum is said to be the oldest continually inhabited place in Africa. Though it is many days' walk inland it was the capital of a maritime empire that extended to Arabia and across the Indian Ocean. However, not much is written or known of the extensive naval voyages by the Axumite kings from their port

The great fallen obelisk of Axum; even pieces of it are massive!

of Adulis on the Red Sea out into the Indian Ocean. One would imagine as well that ships from the Persian Gulf, India, Sri Lanka and Indonesia would have arrived at the port of Adulis. Other voyages would have gone along the eastern African coast.

Says Wikipedia about Axum and the Axumite dynasty:

> Axum was a naval and trading power that ruled the region from about 400 BC into the 10th century. In 1980 UNESCO added Axum's archaeological sites to its list of World Heritage Sites due to their historic value. Located in the Mehakelegnaw Zone of the Tigray Region near the base of the Adwa mountains, Axum has an elevation of 2,131 meters (6,991 ft).

> Axum was the center of the marine trading power known as the Axumite Kingdom, which predated the earliest mentions in Roman era writings. Around 356, its ruler was converted to Christianity by Frumentius. Later, under the reign of Kaleb, Axum was a quasi-ally of Byzantium against the Persian Empire. The historical record is unclear, with ancient church records the primary contemporary sources.

> It is believed it began a long slow decline after the 7th century due partly to the Persians (Zoroastrian) and finally the Arabs contesting old Red Sea trade routes. Eventually Axum was cut off from its principal markets in Alexandria, Byzantium and Southern Europe and its trade share was captured by Arab traders of the era. The Kingdom of Axum was finally destroyed by Gudit, and eventually some of the people of Axum were forced south and their civilization declined. As the kingdom's power declined so did the influence of the city, which is believed to have lost population in the decline, similar to Rome and other cities thrust away from the flow of world events. The last known (nominal) king to reign was crowned in about the 10th century, but the kingdom's influence and power ended long before that.

> ...The Kingdom of Axum had its own written language, Ge'ez, and developed a distinctive architecture exemplified by giant obelisks, the oldest of which (though much smaller) date from 5000–2000 BC. The kingdom was at its height under King Ezana, baptized as Abreha, in the 4th century (which was also when it officially embraced Christianity).

> The Ethiopian Orthodox Church claims that the Church

An old print of the tallest of the standing obelisks at Axum.

of Our Lady Mary of Zion in Axum houses the Biblical Ark of the Covenant, in which lie the Tablets of Law upon which the Ten Commandments are inscribed. The historical records and Ethiopian traditions suggest that it was from Axum that Makeda, the Queen of Sheba, journeyed to visit King Solomon in Jerusalem. She had a son, Menelik, fathered by Solomon. He grew up in Ethiopia but traveled to Jerusalem as a young man to visit his father's homeland. He lived several years in Jerusalem before returning to his country with the Ark of the Covenant. According to the Ethiopian Church and Ethiopian tradition, the Ark still exists in Axum. This same church was the site where Ethiopian emperors were crowned for centuries until the reign of Fasilides, then again beginning with Yohannes IV until the end of the empire. Axum is considered to be the holiest city in Ethiopia and is an important destination of pilgrimages.

So, we already have a problem with the Wikipedia entry on Axum in that Axum wasn't a thriving kingdom with a seaport until around 400 BC but the oldest of the city's many obelisks are said to be from 5000-2000 BC! While Wikipedia says that these older obelisks were much smaller than the later ones, I would disagree and say that the largest obelisks are probably the oldest. The most massive Axum obelisk, the Great Stele, is broken into a number of pieces and once weighed an astonishing 520 tons (estimated)!

It would seem that these huge obelisks are from the same time frame that the large obelisks in Egypt were being erected, and that would be around 2000 BC or earlier. That is a difference of 1600 years in the founding of the Axumite Empire and the construction of its earliest monuments—and they are gigantic in size. It would seem that the Axumite Empire, or at least the city of Axum, was founded long before 400 BC and must easily go back to 3000 or 5000 BC, perhaps earlier. The monuments at Axum are clearly the most stupendous in the whole country of Ethiopia—which has a wealth of amazing buildings and megaliths—and clearly the oldest. They are so old that there is no history as to the erecting of these obelisks—at least the oldest ones—and how they were cut, dressed, moved and erected is a complete mystery to archeologists. Many of them envision elephants helping to haul these massive stone towers down the road to their erection site. The quarry is about four miles west of the obelisk park in the center of the city.

Graham Hancock has this interesting interchange with the high priest of the church that supposedly holds the Ark in his book *The Sign and the Seal:*

> "How powerful?" I asked. "What do you mean?"
>
> The guardian's posture stiffened and he seemed suddenly to grow more alert. There was a pause. Then he chuckled and put a question to me: "Have you seen the stelae?"
>
> "Yes," I replied, "I have seen them."
>
> "How do you think they were raised up?"
>
> I confessed I did not know.
>
> "The Ark was used," whispered the monk darkly, "the Ark and the celestial fire. Men alone could never have done such a thing."[4]

Indeed, the gigantic obelisks (stelae) at Axum are as impressive as any in Egypt and it would seem that Axum was some sort of Egyptian satellite city, and the same engineers, stonemasons and architects who were building the megalithic structures in Egypt were doing the same in Axum. Lake Tana is the source of the Blue Nile and has papyrus growing in it that is made into reed boats, same as in Egypt. Axum is further north than Lake Tana but sits near the Takazze River which feeds into the Atbara River, which meets the Nile north of Khartoum. Certainly there was a strong trade connection with Egypt and the Nile, and it is easy to see how Axum and Lake Tana could once have been early colonies of Egypt.

An Anti-Gravity Machine at Axum?

Let us look at the obelisks that are found at Axum.

The Great Stele is 33 meters long, 3.84 meters wide, 2.35 meters deep, weighing 520 tons.

The Obelisk of Axum, also called the Rome Stele (24.6 meters high, 2.32 meters wide, 1.36 meters deep, weighing 170 tons) had fallen and broken into three pieces (supposedly in the 4th century AD). It was removed by the Italian army in 1937, returned to Ethiopia in 2005, and reinstalled July 31, 2008.

The next tallest is King Ezana's Stele. It is 20.6 meters high above the front baseplate, 2.65 meters wide, 1.18 meters deep, and weighs 160 tons.

Three more stelae measure as follows: 18.2 meters high, 1.56 meters wide, 0.76 meters deep, weighing 56 tons; 15.8 meters high,

198

Electric devices on the walls of the Dendera Temple in Egypt.

2.35 meters wide, 1 meter deep, weighing 75 tons; 15.3 meters high, 1.47 meters wide, 0.78 meters deep, weighing 43 tons.

Standing in the Northern Stelae Field in 2014, I gazed in amazement at the Great Stele, thought to be the largest single block of stone ever erected. For some reason archeologists seem to believe it fell and broke during construction. I could not really see why this would be, and it seems just as likely that it stood erect for hundreds or even thousands of years and then fell, maybe in an earthquake around 400 BC.

When this gigantic 520-ton obelisk fell, it collided with the massive 360-ton slab of stone that is the ceiling of the central chamber of what is called Nefas Mawcha's tomb. This shattered the upper portion of the obelisk and collapsed the tomb's central chamber, scattering the huge roof supports about like broken twigs.

Says the Lonely Planet guide to Ethiopia:

> The megalithic Tomb of Nefas Mawcha consists of a large rectangular central chamber surrounded on three sides by a passage. The tomb is unusual for its large size, the sophistication of the structure and size of the stones used for its construction (the stone that roofs the central chamber measures 17.3 m by 6.4 m and weighs some 360 tons!). The force of the Great Stele crashing into its roof caused the tomb's spectacular collapse.
>
> Locals believe that under this tomb is a 'magic machine,'

199

the original implement the Axumites used to melt stone in order to shape the stelae and tombs. The same type of machine was apparently also used to create some of the rock-hewn churches of Tigray.

This magic machine is pretty interesting, and there seems to be a parallel to what Graham Hancock was told in 1983 by the high priest of the church that holds the Ark.

Clearly this priest believed that the Ark had the power to use celestial fire to cut and lift megaliths, and would also believe that the Ark was brought here circa 950 BC by Menelik. Therefore it would seem that this priest, and others, believed that the obelisks were erected at this time. Was there some sort of anti-gravity machine beneath the massive granite slabs of the so-called Tomb of Nefas Mawcha?

Parts of the Great Stele are so huge that the tourist path at one point goes under the main section of the obelisk that lies at an angle on the partially excavated slope. In fact, very little archeological work and excavation have been done here and it is thought that some obelisks lie buried in the park.

Curiously, the mainstream dating for this huge obelisk and the gigantic collapsed structure called Nefas Mawcha's tomb is the late 4th century AD, near the time of the conversion to Christianity of much of Ethiopia.

This massive block had to be quarried and dressed on the mountain to the west of town and then somehow dragged here to central Axum. It then had to be stood up and aligned vertically to the ground. How this was done is not known and there are no records of the raising of any of the obelisks, though some are ascribed to certain ancient kings. The Great Stele is sometimes referred as King Ramhai's Stele. An amazing structure at the northernmost corner of the Northern

A Djed column holding an orb.

Stelae Field is known locally as the Tomb of King Ramhai, but is usually called the Tomb of the False Door. There, a large stone slab is carved with a false door very reminiscent of the many such carvings in Egyptian temples and tombs.

It would seem that the obelisks are from the time when the obelisks of Egypt were being erected, circa 2000 BC. Over time, burials were made around the obelisks and therefore today they are mistakenly associated with burials. It is thought that somehow each stele is there to mark a tomb. This is far from obvious, and in Egypt obelisks and stelae are not associated with burials at all but are boundary markers and monuments—supposedly representing a ray of the sun.

The Legend of King Ezana

None of the Ethiopian obelisks has any inscriptions, but most have some articulation and carving on them. It seems at the time that Christianity was being introduced into Ethiopia, circa 350 AD, several of the obelisks either fell down or were purposely pushed over. It is said that King Ezana (c.321-360 AD) introduced Christianity to Ethiopia. He was influenced to become a Christian by his childhood tutor Frumentius, a Christian monk from Syria. Frumentius became the first bishop and founder of the Ethiopian Orthodox Church. At that time the Axumite kings ruled over Ethiopia, Eritrea, Yemen, Southern Saudi Arabia, northern Somalia, northern Sudan and southern Egypt. It was a mighty empire that spanned both sides of the Red Sea. They worshiped the sun and moon and a god they named Astar. It is also acknowledged that there was a large Jewish community within the empire.

King Ezana is said to have banished the pagan practice of erecting burial stelae and they became neglected. Over time, many of these stelae fell to the ground, but one of the large obelisks remained standing. It was called the King Ezana Stele, however it would seem he did not erect it. This stele is the centerpiece of the famous painting "Sight of Axum" by Henry Salt (1780–1827).

So it would seem that modern archeologists have fallen into a familiar trap: mistaking the time of some megalithic monuments' destruction with the time that they were built. What modern archeologists are suggesting is that these massive and intricately-carved monoliths were erected in a short time during the beginning of King Ezana's reign with two of the three largest obelisks immediately falling down. And, even more incredibly, shortly after

A depicition of baboons and priests worshipping a Djed column and orb.

erecting his own giant monolith—God knows how—he banned the whole practice and said it was against his religion.

It is painfully obvious to me that most of the stelae in the park are the oldest megaliths in Ethiopia and that King Ezana had nothing to do with the erection of any them. Rather, he may have been responsible for having some of them toppled over. King Ezana issued a number of coins but none of them depict an obelisk—which seems strange if he had erected one of these gigantic 200-ton monsters, a feat of engineering as amazing as any in the ancient world.

There is no written record of King Ezana erecting any stelae and the *Kebra Nagast* makes no mention of the erection of any objects, as great a feat as it would have been. The *Kebra Nagast* speaks of all sorts of amazing and unbelievable things like the flying wagon of Zion, but it does not mention King Ezana and his obelisks.

Modern archeologists admit that they don't know how obelisks were transported and lifted into place. They also admit that it must have taken a pretty amazing effort to do this. They theorize thousands of workers and thousands of meters of rope and rolling logs or sleds and such. The whole thing, it would seem, was a massive undertaking that required hammering these huge objects from the rock, using saws and chisels to square and engrave them, and then some apparatus to move them for several miles to the site. They would then have to have been stood up, either by building a huge crane and hauling them up with pulleys (which is what the Romans did with the Egyptian obelisks they took back to Rome circa 200 BC) or the people might have created mounds to drag the obelisks onto and then tip them down to the ground and use rope to stand them completely upright. Either way some large scaffolding would have to have been erected, even after they were already standing up.

The whole thing is a difficult and, frankly, baffling task. We don't know why anyone would go through the difficult process of quarrying and erecting an obelisk, and this is the same with Egyptian obelisks. Egyptologists have no good explanation for the penchant of Egyptians in the early dynasties to erect obelisks. They are said to be monuments to Ra the sun god, and the typically 300-ton obelisks represent a ray of the sun, giver of life.

Obelisks were also erected at Carnac in Brittany, and the Devil's Arrows of Yorkshire are a series of three obelisks. Obelisks like this have been found in Morocco and Algeria, and it is believed by some archeologists, such as Pierre Honore, that Tiwanaku and Puma Punku in Boliva had obelisks, now fallen and broken.

Erecting obelisks is very tricky and making the very tall ones stand up securely on their bases must have been an engineering difficulty only for the most skillful. Modern archeologists rarely experiment with erecting an obelisk. As we have seen, a PBS television crew for a show called "Nova" tried to erect a small obelisk in Egypt with levers and counterbalances and a cage, but were unable to stand it up. Later a different technique was used in the USA and a small obelisk was successfully raised by hauling it onto a mound of sand, about half as high as the obelisk, and then digging out the sand beneath the base and tipping the obelisk in to a small hole in the ground and then standing it up.

We do not know how the Egyptians raised their obelisks. Despite a wealth of paintings and inscriptions from that culture, the erecting of obelisks is never described. The Ethiopians believe that their obelisks were raised with some sort of anti-gravity power of the Ark of the Covenant. As I looked at the huge stones around me at the Axum obelisk park, I could honestly believe that something like that may indeed have happened.

Obelisks are one of the great mysteries in archeology, and the two countries most known for their obelisks are Egypt and Ethiopia. One would think that the Egyptian obelisks are the largest, but the Great Stele of Axum is the largest known obelisk ever erected. Even if it fell while being erected, it must have stood tall for a few moments, long enough to fall from its great height and break the megalithic slab structure near it as it shattered into five gigantic pieces.

The fall of this obelisk must have been an awesome sight! It certainly highlights the difficulty of raising an obelisk and keeping it erect. Obelisks in Egypt have been standing for over 4,000 years and will probably stand for another 4,000 years. Ethiopia is a good indication that obelisks were not just in Egypt but were in other regions as well.

In fact Ethiopia can boast the fact that they erected the largest obelisk to stand on its narrow base and tower majestically to the sky, no matter how short a few years it might have stood. I suspect that the great Obelisk of Axum stood for many thousands of years, as did the Egyptian obelisks, and was toppled in an earthquake at a later date. Only the Unfinished Obelisk at Aswan would have been larger if it had been completed. Were obelisks once in place around the globe? Is it possible that some obelisks may even be submerged in lakes or oceans? Let us continue our voyage in search of obelisks around the world.

Chapter 6

Obelisks in Europe and Asia

The Ancient Masters were subtle,
mysterious, profound, responsive.
The depth of their knowledge is unfathomable.
—Lao Tzu, *Tao Te Ching* (Chapter 15)

Assyrian Obelisks

We tend to think of Egypt and Ethiopia as the places where obelisks were erected in large numbers, but obelisks can also be found in the Middle East and Europe. Smaller obelisk monuments were known to have been erected by the Assyrians in areas that are currently in Iraq. The Assyrians erected these small obelisks as public monuments that commemorated the achievements of the Assyrian king.

Says Wikipedia:

The British Museum possesses four Assyrian obelisks:

The White Obelisk of Ashurnasirpal I (named due to its color), was discovered by Hormuzd Rassam in 1853 at Nineveh. The obelisk was erected by either Ashurnasirpal I (1050–1031 BC) or Ashurnasirpal II (883–859 BC). The obelisk bears an inscription that refers to the king's seizure of goods, people and herds, which he carried back to the city of Ashur. The reliefs of the Obelisk depict military campaigns, hunting, victory banquets and scenes of tribute bearing.

The Rassam Obelisk, named after its discoverer Hormuzd Rassam, was found on the citadel of Nimrud (ancient Kalhu). It was erected by Ashurnasirpal II, though only survives in fragments. The surviving parts of the reliefs depict scenes of tribute bearing to the king from Syria and the west.

The Black Obelisk was discovered by Sir Austen Henry Layard in 1846 on the citadel of Kalhu. The obelisk was erected by Shalmaneser III and the reliefs depict scenes of tribute bearing as well as the depiction of two subdued rulers, Jehu the Israelite and Sua the Gilzanean, giving gestures of submission to the king. The reliefs on the obelisk have accompanying epigraphs, but besides these the obelisk also possesses a longer inscription that records one of the latest versions of Shalmaneser III's annals, covering the period from his accessional year to his 33rd regnal year.

The Broken Obelisk was also discovered by Hormuzd Rassam at Nineveh. Only the top of this monolith has been reconstructed in the British Museum. The obelisk is the oldest recorded obelisk from Assyria, dating to the 11th century BC.

These small obelisks were essentially monuments to commemorate the achievements of a king and were clearly different in form and function from Egyptian obelisks and other obelisk-type standing stones that we will discuss shortly. Basically the Assyrian obelisks are imitations of true obelisks, which are tall and slender as well as monolithic.

Lebanon has the Temple of the Obelisks that is located at Byblos, north of Beirut. The Temple of Obelisks is an "L" shaped temple with about 25 small obelisks, the tallest being two and half meters. The temple is thought to have been active from 1900 BC through 1600 BC and was used by the Phoenicians.

A hieroglyphic inscription on one of the monuments—the King Abi Shemou Obelisk in the National Museum of Beirut—mentions the god Reshef, god of war and also god of nature and fertility, to whom the temple was probably dedicated. The temple had niches for statuettes, basins and jars for cleansing and sacrifices, but a group of small obelisks were the central part of the temple. In some ways these obelisks were phallic symbols and were similar to the lingam stone of Hindu India. Such phallus lingam stones are often dedicated to the Hindu god Shiva. Ancient Lebanon had a strong relationship with Egypt and Egyptian ships were built from Lebanese cedar, from which a perfume oil is also extracted that was used in the Egyptian mummification process.

There is a Roman obelisk at the ruins of the hippodrome for

The "White Obelisk of Ashurnasirpal I" now at the British Museum. **207**

A small obelisk at the Lebanese National Museum in Beirut.

chariot racing at Tyre, south of Beirut. The obelisk here was found in the underground of the northern end of the hippodrome, and was re-erected in the early 1970s according to obelisks.org. The Japanese-run website says that the obelisk had fallen because of a large earthquake in 502 AD, or another one in 551 AD, which caused widespread destruction around the city of Tyre. There is no inscription and the upper section of the obelisk is missing. The

A coin from Byblos that may depict an obelisk.

obelisk is made of red granite and was probably erected by the Romans between 80 and 200 AD. They probably brought it to Tyre from Egypt. As we will see shortly, the Romans often erected obelisks at one end of the center median at a hippodrome as an object for chariots to turn around.

An obelisk or menhir near Tangiers in Morocco.

209

The broken obelisk at Tyre in Lebanon, once part of a hippodrome.

Roman Obelisks in Europe and Asia

As was the case at Tyre, Lebanon, the Romans erected a number of obelisks in Europe and Asia, often at a hippodrome. Says Wikipedia about Roman obelisks set up in Asia:

> Not all the Egyptian obelisks in the Roman Empire were set up at Rome. Herod the Great imitated his Roman patrons and set up a red granite Egyptian obelisk in the hippodrome of his new city Caesarea in northern Judea. This one is about 40 feet (12 m) tall and weighs about 100 metric tons (110 short tons).
>
> It was discovered by archaeologists and has been re-erected at its former site. In Constantinople, the Eastern Emperor Theodosius shipped an obelisk in AD 390 and had it set up in his hippodrome, where it has weathered Crusaders and Seljuks and stands in the Hippodrome square in modern Istanbul. This one stood 95 feet (29 m) tall and weighed 380 metric tons (420 short tons). Its lower half reputedly also once stood in Istanbul but is now lost. The Istanbul obelisk is 65 feet (20 m) tall.

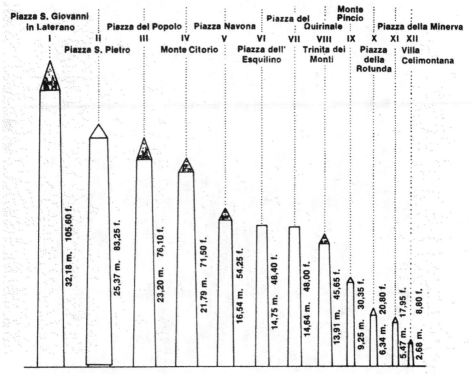

A comparison of the many obelisks now in Rome from Habachi.

The Encyclopedia Britannica tells us that obelisks were used by Phoenicians presumably in what is today Lebanon and possibly Cyprus, and the Romans erected a number in Italy including one of the largest still standing:

> Other peoples, including the Phoenicians and the Canaanites, produced obelisks after Egyptian models, although not generally carved from a single block of stone.
>
> During the time of the Roman emperors, many obelisks were transported from Egypt to what is now Italy. At least a dozen went to the city of Rome itself, including one now in the Piazza San Giovanni in Laterano that was originally erected by Thutmose III (reigned 1479–1426 bce) at Karnak. With a height of 105 feet (32 meters) and a square base with sides of 9 feet (2.7 meters) that tapers to a square top with sides of 6 feet 2 inches (1.88 meters), it weighs approximately 230 tons and is the largest ancient obelisk extant.

Currently, Rome has more obelisks than any city in the world. The most famous is the 25 meter (82 feet), 331-metric-ton (365-short-

An old print of the erection on an obelisk in Rome in 1586 AD.

ton) obelisk at Saint Peter's Square in Rome. The obelisk had been placed in 37 AD next to a wall of the Circus of Nero, flanking what is now St. Peter's Basilica. Pliny the Elder in his book *Natural History* says that the obelisk was transported to Rome from Egypt by order of the Emperor Gaius (Caligula) in a massive barge. The barge had a huge mast of fir wood that four men's arms could not encircle. One hundred and twenty bushels of lentils were used for ballast. Once it had fulfilled its purpose, the massive vessel was no longer wanted, so it was filled with stones and cement and sunk to form the foundations of a new quay at the harbor at the port of Ostia.

But a thousand years later this obelisk would have to be re-erected. It had been moved by the genius of the engineers of the Empire and erected at Rome using wooden erection towers, pulleys, ropes, ramps and cages. It had fallen at the end of the Roman empire, circa 500 AD and had lain as a massive granite monolith on the ground for a thousand years until about 1580 AD when Pope Sixtus V was determined to have it re-erected, and commissioned Michelangelo to move it and place it upright once again. Says Wikipedia on the difficult re-erection:

Re-erecting the obelisk had daunted even Michelangelo, but Sixtus V was determined to erect it in front of St Peter's, of which the nave was yet to be built. He had a full-sized wooden mock-up erected within months of his election. Domenico Fontana, the assistant of Giacomo Della Porta in the Basilica's construction, presented the Pope with a little model crane of wood and a heavy little obelisk of lead, which Sixtus himself was able to raise by turning a little winch with his finger. Fontana was given the project.

The obelisk, half-buried in the debris of the ages, was first excavated as it stood; then it took from 30 April to 17 May 1586 to move it on rollers to the Piazza: it required nearly 1000 men, 140 carthorses, and 47 cranes. The re-erection, scheduled for 14 September, the Feast of the Exaltation of the Cross, was watched by a large crowd. It was a famous feat of engineering, which made the reputation of Fontana, who detailed it in a book illustrated with copperplate etchings, *Della Trasportatione dell'Obelisco Vaticano et delle Fabriche di Nostro Signore Papa Sisto V* (1590), which itself set a new standard in communicating technical information and influenced subsequent architectural

213

publications by its meticulous precision. Before being re-erected the obelisk was exorcised. It is said that Fontana had teams of relay horses to make his getaway if the enterprise failed. When Carlo Maderno came to build the Basilica's nave, he had to put the slightest kink in its axis, to line it precisely with the obelisk.

Three more obelisks were erected in Rome under Sixtus V: the one behind Santa Maria Maggiore (1587), the giant obelisk at the Lateran Basilica (1588), and the one at Piazza del Popolo (1589).

An obelisk stands in front of the church of Trinità dei Monti, at the head of the Spanish Steps. Another obelisk in Rome is sculpted as carried on the back of an elephant. Rome lost one of its obelisks, the Boboli obelisk which had decorated the temple of Isis, where it was uncovered in the 16th century. The Medici claimed it for the Villa Medici, but in 1790 they moved it to the Boboli Gardens attached to the Palazzo Pitti in Florence, and left a replica in its stead.

Several more Egyptian obelisks have been re-erected elsewhere. The best-known examples outside Rome are the pair of 21-meter (69 ft) 187-metric-ton (206-short-ton) Cleopatra's Needles in London (21 meters or 69 feet) and New York City (21 meters or 70 feet) and the 23-meter (75 ft) 227-metric-ton (250-short-ton) obelisk at the Place de la Concorde in Paris.

Other Roman Obelisks

The Romans not only brought obelisks from Egypt but they also commissioned obelisks in an ancient Egyptian style. The Romans created several obelisks in Benevento, Italy and one at Arles, France. There is also the obelisk of Titus Sextius Africanus now in Munich, Germany, but originally found

The obelisk in the Pincio Gardens, Rome.

214

in Rome where it was erected circa 50 AD. Originally from Egypt, it is a monolithic piece of rose granite that is 19 feet (5.8 meters) tall.

The Arles Obelisk, or The Obélisque d'Arles, is a 4th-century Roman obelisk, erected in the center of the Place de la République, in front of the town hall of Arles, a town in southern France. During the time of ancient Rome, Arles was a Roman base to enter Gallia

The curious obelisk at Arles, France, probably from Roman times. **215**

by going upstream on the Rhone River. After Gallia came under Roman rule, Arles prospered as a trader town with Gallia to the north. Because of its Roman history, ancient ruins remain, including an amphitheater from about 80 AD and a Roman Bath from the 4th century AD, which are both well-preserved. Even today the amphitheater is used for concerts and other events. Arles is also the town depicted in many Van Gogh paintings, including his famous "Café Terrace at Night."

The obelisk has a distinctive shape that narrows toward the tip of the upper part. It is also called the Needle of Arles because of that form. There are no inscriptions or hieroglyphs on the monolith and it is thought to be an original obelisk of Roman design rather than one "stolen" from Egypt. The stone was once thought to have been quarried in Europe because it was considered to be "blue porphyry" which is common in Europe. Wikipedia says that the stone is "granite from Asia Minor" but the site obelisks.org says that recent research suggests that stone is actually red granite from Aswan in Upper Egypt. This mysterious obelisk may be from Egypt but quarried and transported by the Romans—we just don't know.

Exactly how it came to be in Arles is apparently open to some debate. According to obelisks.org the obelisk was originally set up in around 80 AD during the era of Titus Flavius Domitianus in the Cirque Romain (Roman circus—a hippodrome for chariot racing) of Arles as a turning point sign on the "spina" (median strip). Wikipedia, however, says that the obelisk was first erected under the Roman emperor Constantine II (circa 340 AD). After the circus was

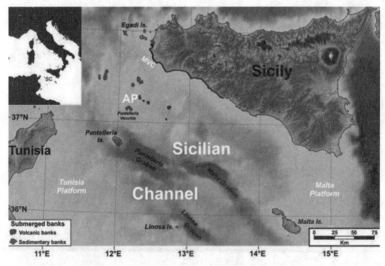

A map showing the location of the sunken obelisk off Sicily.

abandoned in the 6th century, the obelisk fell down and was broken in two parts. It was rediscovered in the 14th century and re-erected on top of a pedestal and then surmounted by a bronze globe and sun on March 26, 1676.

It is interesting to ponder other lost Roman obelisks, fallen and broken over time, that may have been used as turning points on the spina of other hippodromes. We know that other Roman circuses in the Middle East had small obelisks as turning points on the spina as well.

While the Arles Obelisk is a curious monument from Roman times, other obelisks, such as those in northwest France, are much older and much larger than those made during Roman times.

A 9,000-Year-Old Obelisk Found in the Mediterranean

It was reported by Science News (www.sci-news.com) on August 12, 2015 that a 9,000-year-old obelisk was found in the Mediterranean. Said the headline subtitle: "A submerged, 39-foot-long (12 m) monolith has been discovered in the waters off the coast of Sicily at a depth of about 130 feet (40 m)."

Continued the story:

> The man-made monolith is at least 9,350 years old. It weighs about 15 tons and is broken into two parts. It has three regular holes of similar diameter: one that crosses it completely on its top, and another two at two sides of the monolith.
>
> There are no reasonable known natural processes that may produce these elements, according to Dr Emanuele Lodolo from the National Institute of Oceanography and

Photos on the Internet of the sunken obelisk off Sicily.

217

Experimental Geophysics in Italy and Dr Zvi Ben-Avraham from the University of Haifa and Tel Aviv University in Israel, who found the monolith.

"The monolith is made from stone other than those which constitute all the neighboring outcrops, and is quite isolated with respect to them," the scientists said. "It is composed of calcirudites of Late Pleistocene age, as determined from radiocarbon measurements conducted on several shell fragments extracted from the rock samples."

The monolith was found on the Pantelleria Vecchia Bank, a former island of the Sicilian Channel. The island, according to the archaeologists, was dramatically submerged during a flood around 9,300 years ago.

"The obtained age falls chronologically within the beginning of the Mesolithic period of the SE Europe and Middle East," Dr Lodolo and Dr Ben-Avraham said.

"The discovery of the submerged site in the Sicilian Channel may significantly expand our knowledge of the earliest civilizations in the Mediterranean basin and our views on technological innovation and development achieved by the Mesolithic inhabitants."

The monolith required cutting, extraction, transportation and installation, which undoubtedly reveals important technical skills and great engineering. "The belief that our ancestors lacked the knowledge, skill and technology to exploit marine resources or make sea crossings, must be progressively abandoned," the archaeologists said.

"The recent findings of submerged archaeology have definitively removed the idea of technological primitivism

A map showing the location of the sunken obelisk off Sicily.

often attributed to hunter-gatherers and coastal settlers."

The find is described in a paper published July 15 in the *Journal of Archaeological Science: Reports.*

This curious discovery seems to indicate that over 9,000 years ago the Mediterranean was much lower and consisted of dry land around several large lakes. This land contained cities, giant walls and obelisks. These standing stones—obelisks—were probably all over the Mediterranean and are now fallen and submerged.

It is interesting to theorize that there were obelisks on Malta, Gozo and other islands in the past and speculate as to whether there might be sunken monoliths in the water around these islands. The photos from the 2015 paper are quite fascinating and leave no doubt the stones are man-made and many thousands of years old. How many more obelisks are lying undiscovered on the floors of our many seas and oceans?

Obelisks in Brittany

In September of 2018 I led a World Explorers Club group to investigate the obelisks and other megaliths of Brittany including fascinating alignments that extend for miles. Our group drove through the province of Brittany and finally arrived at the small tourist seaside town of Quiberon, our base for the exploration of the megaliths. These megaliths in Brittany have their own myths. Local tradition claims that the reason the stones stand in such straight lines is that they are a Roman legion turned to stone by Merlin. A Christian myth says that the menhirs are pagan soldiers who were in pursuit of Pope Cornelius when he turned them to stone. Brittany is a land of megaliths and mystery.

Brittany is the northwest portion of France, covering the western part of what was known as the Roman province of Armorica. Brittany became an independent kingdom until it was united with the Kingdom of France in 1532, governed as a province that was a separate nation under the crown. Brittany was also referred to as Lesser Britain (as opposed to Great Britain), and is considered to be an ancient Celtic country with linguistic and historical ties to Ireland and Britain through the Irish Sea to the north. Brittany is the site of some of the world's oldest standing architecture; it is home to the Barnenez, a Neolithic monument located near Plouezoch that dates to the early Neolithic, about 4800 BC, and is considered to be one of the earliest megalithic monuments in Europe—in fact, one

An old print of the many huge stones of Brittany.

of the oldest man-made structures in the world. It is also home to the Tumulus Saint Michel, the Carnac stones and other dolmens and menhirs, all of which apparently date to the early 5th millennium BC.

According to Wikipedia:

> Brittany has been inhabited by humans since the Lower Paleolithic. The first settlers were Neanderthals. This population was scarce and very similar to the other Neanderthals found in the whole of Western Europe. Their only original feature was a distinct culture, called "Colombanian." One of the oldest hearths in the world has been found in Plouhinec, Finistère. It is 450,000 years old.
>
> Homo sapiens settled in Brittany around 35,000 years ago. They replaced or absorbed the Neanderthals and developed local industries, similar to the Châtelperronian or

to the Magdalenian. After the last glacial period, the warmer climate allowed the area to become heavily wooded. At that time, Brittany was populated by relatively large communities who started to change their lifestyles from a life of hunting and gathering, to become settled farmers. Agriculture was introduced during the 5th millennium BC by migrants from the south and east. However, the Neolithic Revolution in Brittany did not happen due to a radical change of population, but by slow immigration and exchange of skills.

Neolithic Brittany is characterized by important megalithic production, and it is sometimes designated as the "core area" of megalithic culture. The oldest monuments, cairns, were followed by princely tombs and stone rows. The Morbihan département, on the southern coast, comprises a large share of these structures, including the Carnac stones and the Broken Menhir of Er Grah in the Locmariaquer megaliths, the largest single stone erected by Neolithic people.

For the next few days we explored the area around Quiberon, Plouharnel, Auray, Carnac and the Morbihan Gulf.

Our group met up with researcher Howard Crowhurst in Plouharnel, and Howard took us to some dolmens and standing stones.

Howard Crowhurst is originally from England but has been living in Brittany for about 30 years studying the mysterious megaliths found all over the area. Our first stop was the Crucuno dolmen not far from Plouharnel. It is a massive chamber dolmen with an entrance area that Howard said is aligned with the massive artificial hill known as the Tumulus Saint Michel, which was to the southeast. He told us that the capstone of the dolmen weighed about 40 tons, and pointed out that it was balanced on only three points of the massive standing stones beneath it. Howard, who is the author of several books in English including *Carnac: The Alignments* (2011), maintains that, like the Crucuno dolmen, many of the standing stone alignments and dolmens are aligned with the Tumulus Saint Michel which is itself an astronomical marker.

Another British researcher living in France, Kate Masters, was with our group and I discussed with her whether dolmens, such as the Crucuno dolmen, were really tombs as modern archeologists theorize. She expressed her opinion that the dolmens may have

221

played a part in sky burials, wherein the corpse of someone was put on top of the dolmen which served as a large table. When the sky burial was complete—the consumption of the dead by vultures—the remaining bones were then put inside the dolmen to decay further. So in some ways the dolmen was a burial site, but not as archeologists currently believe.

I had recently been reading about sky burials in a book about the Orkney Islands in northern Scotland. Apparently sky burials were common in Scotland, as well as in Tibet, India, Persia and the

An aerial photo of part of the great stone alignments at Carnac in Brittany.

Americas. Kate was familiar with the sky burials in the Orkneys and thought that the burials there were related to those in Brittany, Persia, Tibet and even Gobekli Tepe in Turkey where recent evidence suggests that sky burials were carried out at that site.

An article published by the *Daily Mail* on June 9, 2016 discussed the work of Dr. Rebecca Crozier, based at the University of the Philippines. Crozier, who previously studied at Edinburgh University, is a specialist in human osteology, forensic archaeology and mortuary analysis. She told the *Daily Mail* that the Orkney Islands are home to at least 72 tombs—known as cairns—dating back as far as 4000 BC. Dr. Crozier's study of two ancient tombs at Quanterness and Quoyness involved re-analyzing more than 12,275 bone fragments previously excavated from the tombs on mainland Orkney and Sanday. She concluded that they had been cut by humans in order to facilitate sky burials.

I thought that it was interesting that the dolmens may have had multiple uses, as tables for sky burials, as chambers for bones and as tunnels to take precise alignments that include the rising or setting of the sun or moon. One member of our group also suggested that the chambers inside the dolmens might have been used for birthing rituals involving midwives and early mothers who gave birth with the help of the priestly class involved in science, astronomy, building and medicines.

Howard Crowhurst agreed that the dolmens seemed to have multiple functions just as modern churches do today, with christenings, marriages, funerals and various festivities on special days of the year held holy by this ancient priestly caste. He told us that the dolmens, like the rows of standing stones and huge artificial mounds called tumuli, were part of a massive system of alignments and measurements that precisely calculated the movements of the sun and moon during the year circa 5000 BC.

Crowhurst, in his aforementioned book, presents an analysis of a megalithic site in the light of the geometrical, mathematical, and astronomical principles by which ancient Carnac was constructed. He has proved that megalithic science of 5000 BC was astonishingly advanced. His book gives us proof that the people who built the ancient monuments were able to perfectly synthesize the movements of the sun and moon at specific latitudes. Crowhurst's is a continuation of the work undertaken by Professor Thom in the 1970s, and he has become a leading authority on how the Carnac site was originally conceived and constructed.

We were glad to have Howard around for a day showing us various fascinating standing stone circles, a stone rectangle, alignments and dolmens. Additionally he explained how the various alignments involved thousands of huge standing stones and capstones for dolmens and the inner chambers of the tumuli.

The Carnac Alignments

The next day we drove south from Quiberon and Plouharnel to the Tumulus Saint Michel, an artificial hill constructed between 5000 BC and 3400 BC. It is about 40 feet high and topped with small chapel dedicated to the archangel Michael. It is said that the mound required 35,000 cubic meters (46,000 cubic yards) of stone and earth. It was first excavated in 1862 by René Galles who dug a series of vertical pits and discovered a stone chamber that contained various funerary objects, such as 15 stone chests, pottery, jewelry and more. He also discovered the original stone entrance which Howard Crowhurst says is an alignment tunnel—he also says that the hill is central to sighting standing stones in the Bay of Morbihan. This would include the great obelisk at Er Grah near the Morbihan Bay, which site also contains the Er Grah tumulus and a dolmen called the Table des Marchands. This area was studied by

Some of the stones in the Carnac alignments in Brittany of which there are many.

the British archeologist Alexander Thom and members of his family who decided that the many megaliths and tumuli were for sighting alignments with the setting sun or the moon. More on them below.

Other tumuli in Brittany include Le Rouzic which was excavated between 1900 and 1907 revealing an inner chamber with stone chests. A chapel was built on top in 1663, was rebuilt several times and still stands today. These tumuli all contain a megalithic chamber with a huge capstone weighing 40 tons or more. They are often called "passage graves" because a megalithic stone passage with a stone roof exited the chamber, often toward the summer solstice. While the exits to the passages were usually not covered over with earth in these man-made hills, or "tumuli," the question has been raised whether these megalithic chambers and passages were originally meant to be freestanding, and not covered over by earth at all. It is not clear when they were encased in earth, and is some cases, when observing gaps in the stonework, now underground, it seems clear that light being able to penetrate the gaps would highlight art work or other features of the original stonework. Another tumulus is at Moustoir, known as Er Mané. It is a passage tomb 279 feet (85 meters) long and 115 feet (35 meters) wide, and 16 feet high. There is a dolmen at the west end, and two chambers at the east end plus a 10-foot-high menhir nearby.

Just prior to lunch we visited the Carnac alignments. Entrance was free and there was a nice information center and bookstore. There are three major groups of stone rows at Carnac—Ménec, Kermario and Kerlescan—that probably once formed a single group, but have been split up because stones were removed by early quarrymen to be used in new buildings.

At the Menec Alignments we walked around eleven converging rows of menhirs that stretch for 1,165 by 100 meters (3,822 by 328 feet). There are what Alexander Thom considered to be the remains of stone circles (cromlechs) at either end. According to the tourist office there is a "cromlech containing 71 stone blocks" at the western end and a very ruined cromlech at the eastern end. The largest stones measuring around 4 meters (13 feet) high, are at the wider, western end.

We continued on to a site that is near the coast of the Bay of Morbihan. This bay contains many islands and small peninsulas and it is acknowledged that when the megaliths of Brittany were built much of the bay was dry land, and many dolmens and standing stones are under the water. Our first stop was at the town of Locmariaquer to see the Grand Menhir Brisé, or the broken menhir, also known as the Er Grah menhir. It is a massive stone the size of an Egyptian obelisk that has fallen and broken into four pieces.

To see the Locmariaquer megaliths we had to buy a ticket and enter into a small museum and bookstore before walking into the complex of Neolithic constructions that comprise (as described above) the elaborate Er Grah tumulus passage grave, a dolmen known as the Table des Marchand and "The Broken Menhir of Er Grah," a huge standing stone that is the largest known single block of stone to have been transported and erected by Neolithic people. There is another large standing stone nearby that was erected at the same time.

The Er Grah Obelisk of Carnac

The Grand Menhir Brisé (broken) at Er Grah is said to be the largest menhir in the world and is situated on a promontory near the water. It is essentially an obelisk. The Broken Menhir of Er Grah is thought to have been erected around 4700 BC and is thought to have been broken around 4000 BC. It was originally about 68 feet high (20.6 meters) and weighed about 330 tons. It was quarried from a rocky outcrop located several kilometers away from where it was erected. That such a gigantic stone block was quarried and moved

A photo of the Er Grah obelisk in Brittany, now in four pieces.

An aerial photo of the Er Grah obelisk in Brittany, now in four pieces.

over a considerable distance and then erected is a remarkable feat according to archeologists, who admit that they do not know why this menhir—or others, numbered in the thousands—were quarried, dressed and erected all over Brittany.

This menhir—standing stone—obelisk—is simply the largest (that we know of) of these impressive stones. Archeologists say that the menhir once had a sculpture of a "hatchet-plough" worked faintly into the surface which is now very eroded and difficult to see. How and why these ancient people erected this huge menhir is still a mystery, but another mystery of this neolithic monolith is how it fell and broke. Archeologists and other specialists argue about the techniques used to transport and erect these stones but they all agree that this feat achieved during the Neolithic era is very remarkable. Still, they have no idea why all these stones were quarried but it must have been a massive physical task that involved large numbers of people and some clever engineering.

We do not know what caused the menhir to topple and break into the four pieces that are currently seen at the site. At one time it was believed that the stone had never stood upright and had fallen while it was being erected. French archeologists are now satisfied that the monolith had once been standing, and, in fact, was the final installment in a procession of menhirs erected at the site, the bases of which can still be seen today. Therefore, it did not fall when it was being erected. Says Wikipedia on the subject:

The most popular theory is that the stone was deliberately pulled down and broken. Certainly other menhirs that accompanied it were removed and reused in the construction of tombs and dolmens nearby. However, in recent years, some archaeologists have favored the explanation of an earthquake or tremor, and this theory is supported by a computer model.

A guidebook I bought at the site mentions that it is an enigma that the pieces of the huge stone pillar fell in such a way that they did not fall in the same direction. The larger bottom section of the pillar fell to the north while the three upper pieces of the monolith fell in a line to the southeast. Why did the pieces fall in different directions? It would appear that the tall stone broke apart before hitting the ground, as if it were struck by a missile or a bolt of lightning.

In the short documentary that is shown at the site it briefly shows the obelisk being hit by an earthquake and the stones breaking up during the quake and then falling to the ground in four separate pieces going in different directions. But how is this possible? One would think that if an earthquake occurred the whole obelisk would fall in the same direction and then break apart on impact with the ground. Would an earthquake snap an obelisk into four separate pieces and have them fall in different directions? It seems unlikely.

More likely it would seem that the obelisk was struck by some energy bolt—either a powerful lightning bolt or a kinetic missile strike—that shattered the monolith and caused it to fall. The footing of the standing stone was only about half a meter and the carefully balanced obelisk could easily lose its equilibrium and fall—but why would it break apart before falling as it apparently did? It remains a mystery to this day. What else do archeologists say about the huge standing stone known as the broken menhir?

The problems of moving and erecting such a huge stone are illustrated in an article by the Thoms published in the *Journal for the History of Astronomy* (No. 2, pages 147-160, 1971) entitled "The Astronomical Significance of the Large Carnac Menhirs." The astronomers, Alexander Thom and his son (with his wife), maintained in the article that the obelisk was a lunar sighting stone. They say that the obelisk, or standing stone, was about 68 feet high (67.6 feet is the current "guestimation") when it stood. The two obelisks attributed to Thutmosis III, now in London and New York, are approximately 68.5 feet high each, about the size of the Er Grah monolith. The Thoms write:

> Er Grah, or The Stone of the Fairies, sometimes known as Le Grand Menhir Brisé, is now broken in four pieces which when measured show that the total length must have been at least 67 ft. From its cubic content it is estimated to weigh over 340 tons.
>
> Hulle thinks it came from the Cote Sauvage on the west coast of the Quiberon Peninsula. His suggestion that it was brought round by sea takes no account of the fact that the sea level relative to this coast was definitely lower in Megalithic times; neither does he take into account the fact that a raft of solid timber about 100 x 50 x 4 feet would be necessary— with the menhir submerged. It is not clear how such a raft could be controlled or indeed moved in the tidal waters round the Peninsula.
>
> Assuming that that the stone came by land, a prepared track (? of timber) must have been made for the large rollers necessary and a pull of perhaps 50 tons applied (how?) on the level, unless indeed the rollers were rotated by levers. It took perhaps decades of work and yet there it lies, a mute reminder of the skill, energy and determination of the

A photo of the top portion of the shattered Er Grah obelisk in Brittany.

engineers who erected it more than three thousand years ago.

In Britain we find that the tallest stones are usually lunar backsights, but there seems no need to use a stone of this size as a backsight. If, on the other hand, it was a foresight, the reason for its position and height becomes clear, especially if it was intended as a universal foresight to be used from several directions. There are eight main values to consider, corresponding to the rising and setting of the Moon at the standstills when the declination was plus or minus. ...It has now been shown that there is at least one site on each of the eight lines which has the necessary room for side movement.

We must now try to think of how a position was found for Er Grah which would have satisfied the requirements. Increasingly careful observations of the Moon had probably been made for hundreds of years. These would have revealed unexplained anomalies due to variations in parallax and refraction, and so it may have been considered necessary to observe at the major and minor standstills at both rising and setting. At each standstill there were 10 or 12 lunations when the monthly declination maximum and minimum could be used. At each maximum or minimum, parties would be out at all possible places trying to see the Moon rise or set behind high trial poles. At night these poles would have needed torches at the tops because any other marks would not be visible until actually silhouetted on the Moon's disc. Meantime some earlier existing observatory must have been in use so that erectors could be kept informed about the kind of maximum which was being observed; they would need to know the state of the perturbation.

Then there would ensue the nine years of waiting till the next standstill when the other four sites were being sought. The magnitude of the task was enhanced by the decision to make the same foresight serve both standstills. We can understand why this was considered necessary when we think of the decades of work involved in cutting, shaping, transporting and erecting one suitable foresight. It is evident that whereas some of the sites, such as Quiberon, used the top of the foresight of Er Grah, others, such as Kerran, used the lower portion. This probably militated against the use of a mound with a smaller menhir on the top. Much has rightly been written about the labour of putting Er Grah in position, but a full consideration of the labour of finding the site shows that this may have been a comparable task.

We now know that for a stone 60 ft. high the sighting is perfect. We do not know that all the backsights were completed. But the fact that we have not yet found any trace of a sector to the east does not prove that the eastern sites were not used because the stones may have been removed. Perhaps the extrapolation was done by the simpler triangle method or perhaps it was done at a central site like Petit Menec.

Much of this gigantic astronomical observatory is probably under water. Many of the megaliths along the Brittany coast are apparently submerged. Many famous sites actually lead into the water, and some megaliths can be seen at low tide when they are barely above the surface. Many of the long lines of standing stones at Carnac and around the Morbihan Gulf were apparently built when the geography of Brittany was quite different.

As we have seen, near the town of Carnac is the famous alignment of hundreds of standing stones. They too are apparently part of some huge astronomical observatory. In another article by the Thoms entitled "The Carnac Alignments" for the *Journal for the History of Astronomy* (No. 3, pages 11-26, 1972), they conclude that Carnac is also a lunar observatory of vast proportions. Say the Thoms about the Menec alignments at Carnac:

A remarkable feature is the great accuracy of measurement with which the rows were set out. It cannot be too strongly emphasized that the precision was far greater than could have been achieved by using ropes. The only alternative available to the erectors was to use two measuring rods (of oak or whale bone?). These were probably 6.802 ft. long, shaped on the ends to reduce the error produced by malalignment. Each rod would be rigidly supported to be level but we can only surmise how the engineers dealt with the inevitable—steps—when the ground was not level.

It may be noted that the value for the Megalithic yard found in Britain is 2.720 plus or minus 0.003 ft. and that found above is 2.721 plus or minus 0.001 ft. Such accuracy is today attained only by trained surveyors using good

231

modern equipment. How then did Megalithic Man not only achieve it in one district but carry the unit to other districts separated by greater distances? How was the unit taken, for example, northwards to the Orkney Islands? Certainly not by making copies of copies of copies. There must have been some apparatus for standardizing the rods which almost certainly were issued from a controlling, or at least advising, centre.

The Thoms see Carnac as part of an ancient and huge system that was used over much of Europe. In their article they conclude:

> The organization and administration necessary to build the Breton alignments and erect Er Grah obviously spread over a wide area, but the evidence of the measurements shows that a very much wider area was in close contact with the central control. The geometry of the two egg-shaped cromlechs at Le Menec is identical with that found in British sites. The apices of triangles with integral sides forming the centres for arcs with integral radii are features in common, and on both sides of the Channel the perimeters are multiples of the rod.
>
> The extensive nature of the sites in Brittany may suggest that this was the main centre, but we must not lose sight of the fact that so far none of the Breton sites examined has a geometry comparable with that found at Avebury in complication of design, or in difficulty of layout.
>
> It has been shown elsewhere that the divergent stone rows in Caithness could have been used as ancillary equipment for lunar observations, and in our former paper we have seen that the Petit Menec and St. Pierre sites were probably used in the same way."

The Thoms confess at the end of their article, "We do not know how the main Carnac alignments were used…"

The many megaliths of Brittany are an archeological enigma that French archeologists have yet to solve. When I asked the French archeologist at the Carnac information office while filming a segment for the Ancient Aliens show in 2009, she told me that they did not know who the builders were or why they had placed these many rows of stones and tumuli. How it was done was a matter of conjecture, but the sheer movement and erection of such a large stone would show that the builders knew how to do such feats and probably that it was easy for them to do such amazing work, not difficult.

So, who were these megalith masterminds who were building Carnac circa 4700 BC? Was it the Egyptians or the proto-Egyptians known as the Empire of Osiris? Was the Carnac area a place where obelisks and stone alignments were erected by the thousands by the Osirian-Celts in order to track the sun and the moon from latitudes that were closer to the north pole? Indeed, many archeoastronomical sites seem to be observatories meant to get closer to the north pole such as those in Scotland and Scandinavia.

Carnac likens itself to the important Egyptian Temple of Karnak. The Egyptian Karnak is a huge building which also has long rows of megalithic columns which once supported a huge roof. Were there great processions down the rows of stones at certain times of the year? Were the dolmens and tumuli more than tombs but rather sacred sites where women also gave birth?

It would seem that these people possessed advanced technologies such as diamond saws, wire saws, cranes, pulleys, levers and sophisticated lifting devices. Did they have the powers of levitation and anti-gravity? Was Merlin involved in the Carnac stones as ancient tradition suggested? Was the system of stone monoliths part of an ancient energy system involving granite obelisks and rows of standing stones? The idea is fantastic!

Are there other, even larger menhirs under the water near Carnac? One example of a known submerged megalithic structure is the Covered Alleyway of Kernic in the District of Plousescat, Finistére, now submerged at high tide. Much of the mystery of the megaliths of Brittany probably lies underwater. The many stories of Atlantis, floods and cataclysms seem to be the history of Brittany and the Atlantic beyond. The people of Brittany are people of the sea, but they are also farmers and shepherds and people of the megaliths.

The Devil's Arrows—Obelisks of Yorkshire

My interest in obelisks eventually drove me to England, where a series of huge standing stones in Yorkshire known as the Devil's Arrows had drawn my attention. I took advantage of some time in the United Kingdom to make a visit to this strange site a few years ago. My wife Jennifer and I made the train trip from London to

York, capital of Yorkshire, and the next day we took a local bus to the town of Boroughbridge some 15 miles west of York.

Boroughbridge is a pleasant town, the site of at least one famous battle, allegedly involving a local hero, Robin Hood, in 1322 AD. While most people associate Robin Hood with the Sherwood Forest to the southwest of Yorkshire, according to *The True History of Robin Hood* by J.W. Walker, the famous outlaw began life as Robert Hoode in Wakefield, Yorkshire. He was called to arms when the Earl of Lancaster raised an army to fight against Edward II. When Lancaster's forces were defeated at Boroughbridge his followers who were not killed or captured escaped into the surrounding forest. So it was in the nearby Barnsdale Forest that Hood began his legendary career. A visitor to the region in the 1530s, John Leland, wrote that: "...betwixt Milburne and Feribrigge I saw the woodi and famose forest of Barnsdale wher they say that Robyn Hudde lyvid like an outlaw."

A short walk from the town square brings one to a large field beside the A1 Motorway; there the three towering monoliths stand as they have for an estimated 4,700 years. They are known as the Devil's Arrows, and were in old times known variously as The Devil's Bolts, The Three Greyhounds or The Three Sisters.

According to Ronald Walker of the Boroughbridge Historical Society, writing in the local brochure available at the small tourist information office, "The three huge standing stones on the western outskirts of Boroughbridge are among the least understood and most neglected historic monuments in Britain. Where they came from, how many there were originally, what their purpose is, and who placed them and when, have been for hundreds of years–and are still today–matters of conjecture."

The stones are 18ft, 22ft and 22ft 6in tall, the last of these being taller than anything at Stonehenge (by only a few inches). The smallest of the stones is rectangular – about 8ft 6in by 4ft 6in. The 22ft stone is 5ft by 4ft in girth and the third and tallest 4ft 6in by 4ft. This last and tallest stands by the roadside

Devil's Arrows, Boroughbridge.

An old print of the Devil's Arrows, 1895.

Two of the Devil's Arrows in Yorkshire.

among trees immediately on the west side of Roecliffe Lane, while the other two are across the road in a field. The smallest of the stones is the farthest away from the road.

The stones are quite impressive, and rather than being rough, natural stones that were uprighted, these stones have been dressed and squared, as if they were obelisks. The tops are heavily worn, and deep grooves from the frequent rains descend from the tops of the stones down the sides. The largest stones appear to be nearly square and it is tempting to think that they were all much taller when originally erected.

The story which led to the current name, The Devil's Arrows, is thought to date from the end of the 17th century: Old Nick, irritated by some slight from Aldborough (a small village to the east of Boroughbridge), threw the stones at the village from his stance on How Hill. His aim, or his strength, being below par the "arrows" fell short by a good mile. It was also claimed in the old days (and perhaps now) that walking 12 times around the stones counter-clockwise will raise the Devil.

Walker repeats that it has been suggested that the stones were part of (or perhaps were intended to be part of) an immense henge (circular earthwork) a mile in diameter. Considering the size of the megaliths, this seems pretty amazing, and such a large circular henge must have had a number of giant stones in it, including other

235

obelisks.

Walker says that the most likely theory is that the stones date from around 2000 BC and that they are probably part of an isolated single row of stones—one of a large number of such megalithic monuments scattered through Western Europe.

Walker says that stone rows vary in complexity, "starting with a pair of standing stones and going on to short rows and then longer rows. The long rows, of which the Arrows would seem to have been an early example, are found in south west England and Northern

The deep grooves can be seen on this Devil's Arrow in Yorkshire.

Ireland and seem to follow on from the earlier long double rows or avenues (such as that at Avebury). The stones of these rows do not compare in size with the Arrows."

Still, such stones are unexplained, and since in most cases the stones have been brought from a long distance, the "why" of the location is also a mystery. Plus, it is obvious to all the researchers that it took considerable effort to quarry and move these stones into place, so the "how" question about the Devil's Arrows is particularly mysterious.

Walker says that it is almost certain that there were at least four stones in the row and possibly five or even more. He says that a report by the visitor John Leland (who also wrote about Robin Hood) in the 1530s gives a clear and detailed description of four standing stones. Thirty years later he says that William Camden wrote of seeing "foure huge stones, of pyramidall forme, but very rudely wrought, set as it were in a straight and direct line... whereof one was lately pulled downe by some that hoped, though in vaine, to find treasure."

Walker says that the upper section of the fourth stone is claimed to stand in the grounds of Aldborough Manor and the lower part is believed to form part of the bridge which crosses the River Tutt in St. Helena just a few hundred yards away on the route into the town center. Large pieces of the same millstone grit that comprises the Arrows have turned up in the garden of a house bordering the field in which the enclosure containing the largest arrow stands.

The remaining three arrows stand in a NNW-SSE line almost 200 yards long! Marks left by a wedge in one of them show that attempts have been made to break up these stones, too, for building material. There is little doubt all of the stones were taller than they are now.

As noted above, the stones are composed of millstone grit and Walker says that the likely source is Plumpton Rocks two miles south of Knaresborough where erosion has produced large quantities of individual slabs. This is about nine miles away from Boroughbridge. Since the lightest of the monoliths weighs over 25 tons, it would have been a tremendous effort using

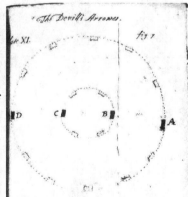

An old plan of the Devil's Arrows.

known technology to haul it to its present location—it would have been difficult, yet possible, even for so-called primitive Neolithic Britains, whoever they were. The big question to me is why? It must have been very important to erect these standing stones—literal obelisks—in the middle of Yorkshire, but archeologists are completely baffled as to the reason.

Walker suggests that the stones could be pulled by teams of 200 men using wood fiber ropes to tow the stone on a sled over a wooden track (which would be picked up and moved as the stone progressed). He says that it is estimated that the arduous pull from Plumpton to Boroughbridge would have taken six months. He says that arriving at the site, "the stone could be raised by dragging it to a prepared hole where it would be slid down a sloping side of the hole and then pulled upright. (It is noticeable that each of the stones inclines slightly to the south)."

This is essentially the same method believed to be used by the Egyptians to erect obelisks. Early archeologists speculated that somehow the Romans had erected the Arrows, but now it is known that the stones were in place much earlier than the Roman occupation. Says Walker:

One of the Devil's Arrows in Yorkshire.

Some people have suggested that the stones were erected by the Romans to commemorate some great victory and there certainly was a Roman fort immediately to the west of the stones. Others have attempted, without much success, to connect them to ley lines. More feasible is the theory that the line was built in prehistoric times to align with the southernmost summer moonrise. Inevitably a religious purpose is ascribed to the stones more than any other. But it has also been suggested that they are monuments to the power or prestige of a local chief, that they are an avenue leading to a henge, or ford, or burial, or...

There is also a suggestion that glacial activity could have carried the stones from the Northallerton area and deposited them only a mile or two from Boroughbridge–which would have considerably simplified the transportation problems.

Walker says that the first recorded excavation at the foot of the stones was in 1709 when a nine-foot area around the central stone was opened. This excavation revealed that cobbles, grit and clay had been packed around the stone to a depth of 5 feet. The base of the stone had been worked to produce a flat bottom that sat squarely on hard packed clay beneath it. The stone below ground had been dressed by pointed tools to give it a smooth appearance. Above ground, of course, weathering has roughened the surface of the stones. Therefore, these stones were originally cut and dressed as if they were square obelisks, including a flat, squared base.

A lead box containing four William III and Queen Anne halfpennies was deposited at the foot of the stone before the hole was re-filled. An excavation of the smallest stone in 1876 resulted in a hole 4 feet 6 inches deep. Five years later, an excavation of the tallest Arrow showed that it was buried 6 feet beneath the ground.

We walked around the site for an hour, tramping through a farmer's wet field to see the two smaller Arrows. We took photos, looked up at the grooves worn into the tops, and wondered at the general massiveness of these standing stones. They were awesome!

As we left Boroughbridge by bus, I realized that we had forgotten to walk around one of the arrows 12 times to try to raise the Devil. As the bus barreled down the winding country roads back to York, I thought that it was probably best that we had not done so. It will remain for others to test this ancient legend—and incur the wrath of

Old Nick or perhaps Ba'al or another undesirable spirit.

Rudston Monolith — an Ancient Obelisk?

My interest in the Devil's Arrows being the possible remains of ancient obelisks led me to another curious standing stone in Yorkshire called the Rudston Monolith. This giant, squared obelisk-like standing stone is almost 26 feet high and stands at the northeast corner of the Rudston Parish Church of All Saints. Rudston is a small village near the seaside town of Bridlington, in the East Riding area of Yorkshire. The Rudston Monolith is made from Moor Grit Conglomerate from the Late Neolithic Period, which is a material that can be found in the Cleveland Hills inland from Whitby, Yorkshire.

Photos of the Rudston Monolith looked very interesting, and it seemed that this ancient standing stone might also have originally been an obelisk, as I suspected the Devil's Arrows were. Jennifer and I took the local train to Scarborough on Yorkshire's eastern coast and then another train south to the coastal tourist town of Bridlington.

We found a Bed and Breakfast near the seashore and went out that night for a dinner of the local fish, followed by an evening at the neighborhood pub, where the patrons were happily engaged in a music quiz led by a 60-something deejay. The next day we discovered that Rudston was only about 14 miles away from Bridlington, but the local bus only went out to the town one day of the week. This day was Thursday, but we were trying to get there on a Sunday. We were forced to hire a taxi to make the trip. It was a brief drive through the pleasant Yorkshire countryside and we were quickly at the tiny village of Rudston.

Rudston has a few houses, a pub, and a now unused church with a graveyard on its north side, wherein sits the massive monolith. Clearly, the

An old print of the Rudston monolith.

240

An old photo of the Rudston monolith and the church.

church had been built at this location hundreds of years ago because of the spectacular monolith that once apparently stood alone on the Rudston hill.

Our taxi parked at a stone wall next to the church and Jennifer and I examined the monolith and walked around the church grounds. The pleasant village church was open, and it was a Sunday, but no one was inside. It was a sunny, early March day around noon, and the English daffodils were just beginning to bloom around the churchyard.

Officially, the Rudston Monolith is the tallest megalith (or standing stone) in the United Kingdom at over 25 feet high (7.6 meters). It has been dressed and squared, and has the appearance of the bottom part of an obelisk. The pointed top has a lead cap to protect against further erosion.

According to the interesting website The Megalithic Portal (www.megalith.co.uk) as much of the Rudston Monolith may be buried in the ground as can be seen above ground. This would make the current stone 50 feet in total length and it must have been initially larger than this, as some of it has definitely eroded or been purposely cut away. Indeed, there is evidence, as in Brittany and other places, that the top of the standing stone had been carved into a Christian cross at some point in the Middle Ages.

The Megalithic Portal also mentions a similar legend to that of the Devil's Arrows, of the Devil throwing the stone at the Rudston church and missing. According to the site, the nearest source (Cayton or Cornelian Bay) of stone of the type that comprises the monolith is 9.9 miles (16 kilometers) north of the site. The website says that archeologists estimate that the monolith was probably erected around 1,600 BC, which is similar to the 2000 BC date

241

estimated for the erection of the Devil's Arrows at Boroughbridge. In my opinion, both the Devil's Arrows and the Rudston Monolith were probably erected at about the same time, around 2,000 BC, if not earlier.

The Megalithic Portal also says that there is one other smaller stone, of the same type, in the churchyard, which was once situated near the large stone. They also say that the Norman church was almost certainly intentionally built on a site that was already considered sacred, a practice which was common through the country. They also say that the name of Rudston may come from the old English "Rood-stane," meaning "cross-stone," implying that a stone already venerated was adapted for Christian purposes.

The Megalithic Portal mentions that in 1861 a Reverend P. Royston did some excavation around the stone. He is quoted in the book *Old Yorkshire Vol. 1,* by William Smith (1891), and states that in 1861 during leveling of the churchyard an additional 1.5 metres (4.9 feet) of the monolith was buried. The weight is estimated at 40 tons (aprox. 40,000 kg).

Sir William Strickland (apparently the 6th Bart of Boynton, 1753-1834) is reported to have conducted the experiment in the late 1700s that determined that there was as much of the stone below ground as is visible above. Strickland found many skulls during his dig and suggests they might have been sacrificial.

The Megalithic Portal says that the top appears to have broken off the stone. If pointed, the stone would originally have stood at

The Rudston monolith stands next to an old church.

least 8.5 meters (28 feet) and probably higher. The site says that in 1773 the stone was capped in lead; this was later removed, though the stone is capped currently.

The site also suggests that fossilized dinosaur footprints on one side of the stone may have contributed to its importance to those who erected it.

Rather than being square like an obelisk (and two of the Devil's Arrows), the stone is fairly slender, with two large flat faces. The flat face of the stone is oriented toward the midwinter sunrise in the southeast.

The site also says that lines across the countryside created by removing soil and grass from above the turf, existed in the area, and have been linked to the stone. These may have inspired the ley line theory that is also applied to the Devil's Arrows. The Megalithic Portal says that there are many other earthworks in the area, including burial mounds and cursuses (more on these later).

Yorkshire historian Mike Thornton supports the information given by The Megalithic Portal and provides further detail in his web pages (www.thornton1.freeserve.co.uk/rudston1.htm):

> Very little is known about this ancient piece of rock in the churchyard at Rudston village near Bridlington. ...Nearby, there are traces of ancient lines that were drawn by removing the grass and exposing the white chalk underneath. There are also burial mounds not far away at Willy Howe, Duggleby Howe and Ba'l Hill. The significance of the standing stone must have been religious. Its date is only roughly estimated as between 2000 and 3000 BC.

Thornton says that in 1872, the local curate mentioned above, Rev. P. Royston, made measurements of the monolith. He reports the results as follows:

> 25 feet 4 inches high
> 6 feet 1 inch wide on the East side
> 5 feet 9 inches wide on the West side
> 2 feet 9 inches thick on the North side
> 2 feet 3 inches thick on the South side
> Royston says the ground was leveled in 1861, burying a further 5 feet of the monolith. Sir William Strickland had previously dug down and shown there was as much below

ground as above before the leveling. This would make the total length about 60 feet. Strickland found great quantities of skulls during his dig and suggested they might have been sacrificial. (*Old Yorkshire Vol. 1*, by William Smith, 1891)

The cursus system, the monolith and several henges and mounds follow the valley of the stream known as the Gypsey Race. This eventually reaches the sea at Bridlington Harbour.

It is interesting to note the curious names of Ba'l Hill and the local stream known as the Gypsey Race. Both would seem to hint at some Egyptian or Eastern Mediterranean origin for the Rudston Monolith and other Neolithic works. Describing something as "gypsy" is actually giving an Egyptian quality to it; it is curious that a stream in Yorkshire was given an Egyptian-type name.

I was unfamiliar with what a cursus was, so I had to do further research. I discovered that a cursus (plural 'cursus' or 'cursuses') is a name given by early British archaeologists such as William Stukeley, who did much of the early work on Stonehenge, to the large parallel lengths of banks with external ditches which they originally thought were early Roman athletic courses, hence the Latin name cursus, meaning "course."

Cursus earthworks are now understood to be earlier Neolithic structures and they are said to represent some of the oldest prehistoric monumental structures of the British Isles. Their function (thought to be ceremonial) is unknown. They range in length from 50 meters to almost 10 kilometers and the distance between the parallel earthworks can be up to 100 meters. Banks at the terminal ends enclose the cursus.

According to Wikipedia, it has been traditionally thought that the cursuses were used as processional routes. Says Wikipedia:

> They are often aligned on and respect the position of pre-existing long barrows and bank barrows and appear to ignore

The Rudston monolith.

244

difficulties in terrain. The Dorset Cursus, the longest known example, crosses a river and three valleys along its course across Cranborne Chase. It has been conjectured that they were used in rituals connected with ancestor worship, that they follow astronomical alignments or that they served as buffer zones between ceremonial and occupation landscapes. More recent studies have reassessed the original interpretation and argued that they were in fact used for ceremonial competitions. Finds of arrowheads at the terminal ends suggest archery and hunting were important to the builders and that the length of the cursus may have reflected its use as a proving ground for young men involving a journey to adulthood. Anthropological parallels exist for this interpretation.

Examples include the four cursus at Rudston in Yorkshire, that at Fornham All Saints in Suffolk, the Cleaven Dyke in Perthshire and the Dorset cursus. A notable example is the Stonehenge Cursus, within sight of the more famous stone circle, on land belonging to The National Trust's Stonehenge Landscape.

Is it possible that the enigmatic cursuses of Great Britain are the remains of canals or dikes that were originally built to bring ships and barges into interior areas of England from existing rivers that were connected to the sea? Such a theory has been applied to the Devil's Dikes near Oxford. Such a canal system, that was sometimes closed at one or both ends, could have been used for transporting large stones such as the Rudston Monolith and the Devil's Arrows. Were the rivers crossed by the cursuses meant to fill them with water? Obviously, the river must enter a cursus at this point and could be partially diverted to fill up the cursus or dike, which would have to be fairly deep.

What is curious about the Devil's Arrows and the Rudston Monolith is that they are not only among the earliest megaliths in Great Britain, but they are the largest—even bigger than Stonehenge.

As I looked at the Rudston Monolith, I wondered if it had been built at approximately the same time as Stonehenge in Wiltshire? Had there once been a megalithic culture, similar to that in ancient Egypt, that built a network of obelisks and stone temples throughout Britain and even throughout Europe? The idea seemed fantastic, yet the evidence was there.

What culture might this have been that was erecting giant stone temples and obelisks across the countryside? Could it have been an offshoot of the great Egyptian empire itself?

Some researchers suggest that obelisks like the Devil's Arrows were originally massive survey markers for a sophisticated, prehistoric survey team. Another suggestion is that they are giant stone acupuncture needles put in the earth to energize it and give crops vitality. A similar explanation has been given for the round towers in Ireland.

Perhaps the Devil's Arrows were part of a global obelisk system of which small parts can still be seen today. Granite obelisks, admittedly, would be better for this purpose then ones made from millstone grit. Other obelisks near Rudston and Yorkshire might be underwater at this point. Maybe divers in the North Sea will discovered an obelisk or two coming up from the ocean floor at some point in the future. No doubt there will be some speculation that the Devil put them there.

Chapter 7

Obelisks in the Americas

That street, like any other, leads to eternity...
All you have to do is follow it in total silence.
It's time. Go Now! Go!
—Don Juan Mateus,
The Fire Within by Carlos Castenada

Obelisks in South America

I first visited Chavin de Huantar (sometimes just called Chavin) in 1990, taking an overnight bus to Huaraz, the skiing capital of Peru. Huaraz is the capital of the Ancash Department and is about 420 kilometers north of Lima. Located at over 10,000 feet (3,000 meters), Huaraz is surrounded by the highest mountains in Peru known as the Cordillera Blanca, or White Range, because of their permanent glaciers and heavy snow covering.

I stayed for a couple of nights in this pleasant alpine town of skiers, mountain climbers and trekkers, and then took a local bus east over a high pass to the town of Chavin. I stayed there for two nights and investigated the archeological site during the day. I was amazed by the complexity of the site at the time, and I returned for another visit in early 2011 with my wife Jennifer.

Chavin de Huantar is thought by many Peruvian archeologists to be the genesis of South American civilization. It is also the site of several obelisks or obelisk-like monuments. Peruvian archeologists think that the early phases of Chavin began over 5,000 years ago, circa 3000 BC and the massive structures that can be seen today were in use before 900 BC.

The building of Tiwanaku followed shortly after the building of Chavin, according to the mainstream view. Indeed, there are many similarities between Chavin and Tiwanaku and it would seem that the two are very much linked. Chavin is thought to have ceased to be a major center around 400 BC and became ruins in a remote and unpopulated mountain valley. As we shall see, obelisks are believed to have been also present at Tiwanaku and Puma Punku near Lake Titicaca as well.

Today, the current village of Chavin, hundreds, even thousands of years old, is largely built of stones salvaged from the old site of Chavin de Huantar. What remains of the archeological site is basically the largest megalithic blocks and the extensive underground structure. Portions have been reconstructed by modern archeologists.

The Tello obelisk, now in the museum, was originally found in a sunken temple excavated by the famous Peruvian

A photo of the Tello Obelisk at Chavin in Peru.

archeologist Julio Tello. Says Encyclopedia Britannica while talking about a sunken courtyard discovered at Chavin:

> Within this court is a square, slightly sunken area, in which was found the Tello obelisk, a rectangular pillar carved in low relief to represent a caiman and covered with Chavín symbolic carvings, such as bands of teeth and animal heads. This is considered to be an object of worship like the Smiling God and Staff God. Carvings found on and around the temple include a cornice of projecting slabs, on the underside of which are carved jaguars, eagles, and snakes, and a number of tenoned heads of men and the Smiling God; they are thought to be decorations or the attendants of gods rather than objects of worship.

A central feature of Chavin de Huantar is the underground temple where an obelisk-like monolith was found carved with the image of a toothy monster said to be the object of worship at the temple. Legends speak of the megalithic complex going nine or 12 stories underground, where a huge gem can be found. Naturally, the complex has long since collapsed and is buried with rubble in the lower parts. Tourists today are allowed to go down to about the third level.

The site is massive and impressive. A good description can be found in the Time-Life book *The Search for El Dorado*:

> Much of Chavin de Huantar's drawing power stemmed from its awe-inspiring architecture and monumental sculptural details. In its early phases the temple consisted of three enormous platform mounds arranged in a U-shape atop a pedestal of cyclopean stone blocks. Gazing upon its massive walls, worshipers beheld a bas-relief frieze of anthropomorphic creatures, spotted jaguars, writhing serpents, and wild birds of prey, below which projected a row of monstrous human heads sculpted from stone blocks weighing as much as half a ton each. Tenons at the back of the heads fit snuggly into mortise joints in the masonry, creating the illusion that the sculptures were floating some

30 feet above the ground. A frieze depicting similar figures — jaguars and exotically costumed humans — adorned the sunken, circular plaza that lay between the three mounds. A white granite staircase climbed from the plaza up to the temple's summit.

...Pilgrims gathering on the New Temple's main plaza would have had a commanding view of the ritual platform. No exterior staircase surmounted this new structure; the mound's apparently inaccessible summit was instead approached by a labyrinthine network of interior passages and stairways. Garbed in fine cotton, woolen, or feathered tunics and adorned with gold nose ornaments, earspools and headdresses, Chavin de Huantar's priests must have elicited considerable wonder when, as if by magic, they emerged on top.

These pilgrims are then thought to have left with various mementos of their visit to the cyclopean complex, including clay objects, gold objects, painted textiles and other "souvenirs." These have been found in distant mountain and coastal areas of Peru. Julio Tello discovered a gold gorget representing two intertwined snakes over 280 miles away from the site (at Chongoyape) that he identified as coming from Chavin de Huantar.[31] Was this gold artifact made at the megalithic site? Were other artifacts, such as the ceramics and fabrics made there as well? Was Chavin a manufacturing center? Archeologists do not really address this issue.

Chavin de Huantar is usually described as a ceremonial center, located at the headwaters of the Maranon River. Two rivers converge at Chavin, the Mosna River and the Huanchecsa River. The builders of the complex actually diverted one of the rivers to enter the complex. Similar things were done at Tiwanaku.

A close-up of the Tello Obelisk.

Chavin is at over 10,000 feet in the Andes and it is in a very steep valley. Why would such a complex, a huge engineering work, be placed in this rather remote and difficult spot?

Well, according to the major archeologists working at the site, it was a religious and ceremonial center "centrally located" to the Andean people and the coastal areas as well. Here the people ingested psychedelic cactus extracts and generally attended religious ceremonies of some kind. While this may be partially true, and certainly such ceremonies with psychotropic substances must have occurred, it seems the builders were very sober geniuses who were designing very complicated facilities. They not only built a pre-designed underground complex with cut and dressed megalithic blocks, but they diverted a river into the complex as part of their design. Why did they do that?

Whenever gigantic blocks of granite, limestone or basalt are being used in a building, it is obvious that a great deal of organized planning is involved. The quarrying of the blocks, dressing them, moving them to the site and then erecting them is a considerable task, not one that should be taken lightly. Whoever the minds behind this endeavor were, they were very smart, skilled in engineering and architecture—and had a forward-thinking mindset. In other words, they were putting a great deal of organizational effort and cost into something that would only pay off later. Chavin was, in my opinion, a mining center where gold, silver, copper and other metals were produced. Obelisks, for whatever purpose, were part of this complex.

The Obelisk at Puma Punku

The German author Pierre Honoré[14] claims that there was an obelisk standing in front of what he calls the Temple of the Moon at Puma Punku within the Tiwanaku complex near the southern shore of Lake Titicaca in Bolivia. The ruins of Tiwanaku and Puma Punku are some of the most astonishing megalithic structures in the world. Today tourists marvel at the fine granite and sandstone stonework that remains at the various archeological sites on the Bolivian side of Lake Titicaca and I always marvel at the huge walls, monolithic doors and heavily articulated H-blocks.

Honoré in his book, *In Search of Quetzalcoatl*[14] (originally published in German as *Quest for the White God*), says:

> The obelisk comes from Egypt, where it was extremely common, and these 'fingers of the sun' used to stand outside the Egyptian temples. Two of the tallest, built by a powerful Pharaoh, are in front of the temple at Heliopolis. Obelisks were known at Crete, for there are two of them shown in frescoes on the sides of the sarcophagus of Hagia Triada; and obelisks also stood outside the temples of Tiahuanaco. In addition, twenty-one human figures in stone, sixteen stelae, and forty-eight sculptures altogether, have been salvaged from the horseshoe-shaped temple of Pucara in Peru. These statues are more rounded than the 'cubist' forms of Tiahuanacan art, but otherwise the two styles tally exactly. The Pucara pottery, too, is like that of Tiahuanco, showing the same winged creatures, the hybrids of man and beast, catlike creatures, hunting trophies and condor heads, fish heads and llamas.
>
> ...On all important sites of ruins some of the great gateways have survived. This is a little surprising, since a gate, the place where a wall has been broken through, might be expected to be a building's weakest part. Three gates have been preserved, or at least can be reconstructed, from three different ancient civilization of Europe and Asia: the Babylonian gate of Nebuchadnezzar, the lion gate at Mycenae and the Hittite gate of Hattusas. All three gates have animals or fabulous monsters as ornaments.
>
> ...There were the remains of more big gates on the great pyramid of Puma Puncu. One of them, the Gate of the Moon, must have been a magnificent sight; it is a monolith similar to the Gate of Sun, also with a frieze in relief, but with fishes instead of the condor and with no winged creatures depicted. Its reliefs still show little boreholes, particularly in deep-set places; these served to fasten plates of silver—the color of the moon—which lined the relief's hollowed-out parts. Puma Puncu was dedicated to the goddess Pachamama, who may have been also a moon goddess.[14]

What Honoré is arguing in his book is that transoceanic contact was occurring between the Americas and the Mediterranean and that obelisks, such as those at Puma Punku (he spells it Puma Puncu), are proof that there is an influence from Egypt and Crete in the ruins found around the mining district of Lake Titicaca. The fact that the early German and Polish archeologists like Arthur Posnansky believed that two obelisks once stood at the "Temple of

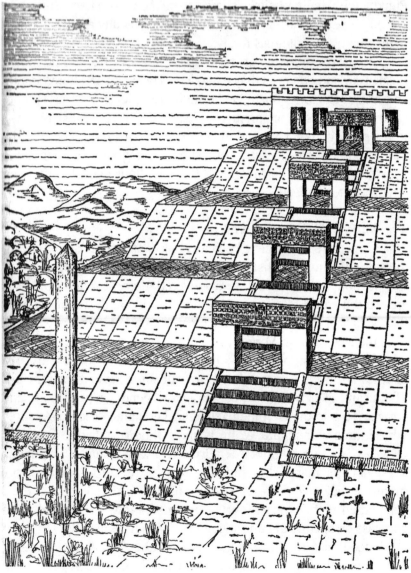

A drawing of the obelisk that once stood at Puma Punku in Bolivia.

253

the Sun" and the "Temple of the Moon" is remarkable because our knowledge of obelisks outside of Egypt and Ethiopia is so limited.

Honoré merely passes over the fact as he discusses Mediterranean influence over cultures in South and Central America. He also briefly discusses the use of bronze in the Lake Titicaca area and that the earliest known bronze artifact comes from the island of Sardinia and is dated to about 1900 BC. He thinks that Tiwanaku and Puma Punku, with their obelisks, dates from approximately this same period, or perhaps around 1200 BC, a date typically agreed upon by mainstream archeologists today.

The broken remains of these fallen obelisks were removed by Posnansky and others as they excavated Tiwanaku and Puma Punku and attempted to reconstruct the main Temple of Sun, as Honoré calls it. Posnansky apparently writes about these fallen obelisks in his early books. Excavation continues to this day at both Puma Punku and Tiwanaku and both sites are marvels of megalithic construction—now destroyed. With or without obelisks, they are amazing sites, indeed.

The Marysburgh Vortex and the Obelisk

While there might have been an obelisk at Tiwanaku in the past, the story of a huge obelisk that once stood in Lake Ontario is one of the most bizarre stories I have discovered in my search for obelisks around the world. A Canadian author named Hugh Cochrane even thinks that this obelisk may have created a vortex in the lake that he calls the Marysburgh Vortex.

The Marysburgh Vortex is mentioned in Wikipedia as an "area of eastern Lake Ontario with a history of shipwrecks during the age of sail and steam that has encouraged legends, superstitions and comparisons to the Bermuda Triangle. The name describes an area whose three corners are Wolfe Island (Ontario), Mexico Bay near Oswego, NY, and Point Petre in Prince Edward County." Wikipedia says that the many shipping disasters are due to "conventional maritime hazards such as bad weather, shifting cargoes, fires, submerged reefs exposed during periods of low water levels, and compass errors due to natural geomagnetic anomalies."

Indeed, natural "geomagnetic anomalies" may have caused some ships to run aground on the rocky shores and sink with all aboard, but Cochrane thinks that an obelisk on the Canadian side of Lake Ontario was the cause of at least one disaster. In his 1980 book, *Gateway to Oblivion*,[20] Cochrane says that there are at least 100 missing vessels, including the *Bavaria* (1889), *George A. Marsh* (1917*)*, *Eliza Quinlan* (1883) and *Star of Suez* (1964). He coined the term "Marysburgh Vortex" and proposed "an unknown invisible vortex of forces" as the cause of the maritime disasters. His book was one of several that came out about "vortex areas" following the publishing success of Charles Berlitz's 1974 book *The Bermuda Triangle*.[19]

Cochrane discusses a number of ship disappearances in the book but the story of an obelisk near Presqu'ile Bay on the north shore of Lake Ontario and the disappearance of a ship named the *Speedy* is probably the most interesting. The story takes place in 1804 with the strange discovery of an obelisk many hundreds of feet tall standing in the lake near the shoreline of Presqu'ile Bay. Says Cochrane in chapter two of his book:

> While the events taking place in this region today appear mysterious and perplexing, one that occurred in the early 1800s would have been classified as astounding. This particular case involved a shipload of government dignitaries and a gigantic stone monolith weighing hundreds of tons. All of these vanished without a trace!
>
> This single most mysterious event took place in Lake Ontario's eastern end in the fall of 1804, when the Canadian government ship *Speedy* vanished along with her crew and passengers.
>
> Present-day records of this event contain the cryptic words "lost on the lake" to explain this disastrous occurrence. But nowhere in these records is there any mention of the mysterious role played by a three-hundred-foot high,

forty-foot-square stone monolith that had been discovered submerged in the lake, and to which this vessel seemed strangely drawn before they both vanished.

In the oft-told versions of the disappearance of the *Speedy*, the story usually begins as the ship prepared to sail from the docks at York (now Toronto) on a Sunday in November 1804. Government officials, representing the majority of the young government of Upper Canada, boarded the ship for the hundred-mile voyage down the lake to Presqu'ile at the eastern end. Among the passengers was an Indian prisoner named Ogetonicut who was being taken to the new town site of Newcastle to be hanged.

After setting sail the *Speedy* made several stops along the lakeshore to pick up other passengers, then the course was set for Presqu'ile Bay. However, the ship was said to have encountered a severe storm and foundered, taking with her all aboard.

That is the simplified version of the saga of the *Speedy* which appears on leaflets and on plaques commemorating this dramatic event. But this version contains nothing about the occurrence that preceded the disappearance of this vessel, or the weird discovery made after the event. These seem to indicate that there were forces at work far more powerful than any government. And they were shaping the history and destiny of this region of North America.

The full story began in the spring of 1804, months before the *Speedy* set out on its fateful voyage. At that time another vessel, the *Lady Murray*, was sailing across the lake toward Presqu'ile Bay on the north side of Lake Ontario. The trip across the lake was uneventful until the ship neared the entrance to the bay. At this point one of the seamen aboard the *Lady Murray* spotted something unusual on the surface of the water. In one small area the waves seemed to be acting strangely in comparison to the water around it.

The seaman brought this anomaly to the attention of the ship's master, Captain Charles Selleck, who also thought it rather peculiar. He ordered the ship stopped and had a boat lowered over the side, and he and some of the crew went to

256

investigate the area at close range.

While the rest of the crew gathered at the rail, Selleck and the others approached cautiously and peered down, seeking the cause of the wave action. At first the answer seemed simple enough. There was a shoal or rock just under the surface. However, a closer examination soon made them change their minds and gave them a healthy respect for that particular area of the lake from then on. What Selleck and his crew discovered was that this "rock" was approximately forty feet square and less than three feet under the surface. But when they sounded around it, they learned that the rock's sides dropped straight down for three hundred feet.

This came as a bit of a shock since it was believed that the water at this point was relatively shallow. This so puzzled Captain Selleck that he ordered the soundings done again. The soundings were accurate. The huge stone monolith was sitting upright like a giant tombstone in an incredible fifty fathoms of water!

The find was astounding, to say the least. In fact it was so unusual that when news of it was released, hundreds of local inhabitants came to the site in small boats just to peer down into the depths and sound the area around the strange object themselves.

When Captain Selleck returned to his ship he dutifully recorded the discovery in his logbook. This monolith was a definite hazard to navigation and, although shipping on the lake was still in its infancy, the authorities and other ship's captains would have to be warned. [20]

This amazing story being told by Cochrane appears to be true and literally hundreds of people rowed out to the monolith in their small boats and witnessed the awesome structure. One would imagine that the obelisk had been placed on dry land when Lake Ontario was smaller and gradually became underwater as the lake filled up with fresh water and rose to its present shoreline and depth. Perhaps an earthquake along the St. Lawrence River had closed up an outlet to the smaller Lake Ontario and it became the larger lake that it is today. Could the obelisk have been placed there sometime after the last ice age, perhaps around 5000 BC or even earlier?

The story of the Marysburgh Vortex and the obelisk does not end with discovery of the monolith near Presqu'ile Bay. Says Cochrane:

> Over the next few months the site drew many visitors who pushed, probed, and jabbed at the huge stone in an attempt to topple it. But it was immovable and solid.
>
> Among the visitors to this site was Captain Thomas Paxton of the government schooner *Speedy*. It was brought to his attention by Captain Selleck, who personally took him to the site to alert him to the danger that the monolith posed for ships approaching the bay. No one recorded what Paxton's reactions were, yet he was destined by fate to become deeply involved with this submerged mystery before many months had passed.
>
> It was around this same time that the Indian named Ogetonicut was accused of killing a white man at a small settlement just north of the lake. He had been living with members of his tribe, who were camped for the summer on one of the islands that form Toronto's bay. When caught, he was taken into custody and held for trial.
>
> However, the trial raised some legal problems. The murder had been committed in another district and the law required that the trial be held there. But there were no suitable court facilities available in that district and no proper center for government activities. This sent the ponderous wheels of bureaucracy into motion and a solution was soon arrived at. A temporary courthouse would be arranged at the site at Presqu'ile Bay. Ogetonicut could then be tried and hanged, and the occasion would serve to establish that location as the new town of Newcastle.
>
> With this decided, arrangements were made for the government schooner *Speedy* to transport Judge Thomas Cochrane, court officials, and a selected group of dignitaries who were to officiate over the establishment of the new town.
>
> Strangely enough, the *Speedy* had two alternate captains. One was Paxton and the other was James Richardson. On

this trip it was Captain Richardson's turn to act as master of the government ship. But it turned out that Richardson had a deep foreboding about the trip and he tried to persuade the officials to put it off. They in turn rejected the idea of postponing what was planned as a gala event. Richardson persisted in his warning and was told to step aside and allow Captain Paxton to command the ship. When his warnings of doom failed, he gave the command to Paxton, then tried to alert some of the others who were to board the ship.

In the end Richardson's warnings were ignored and the ship was prepared for departure.

The passengers who boarded the vessel that day were all destined to meet their fate on that voyage, including two young children, who had been put aboard by their parents and left in the captain's charge. There had not been enough money to purchase passage for the whole family. After exchanging good-byes the parents set out to walk the one hundred miles to Presqu'ile. They never saw their children again. The only unwilling passenger on the *Speedy* was Ogetonicut. He had no choice and was taken aboard in chains.

The stops along the lake were to pick up witnesses to the crime. However, some inner sense had warned the witnesses against the voyage and they traveled overland all night to reach Presqu'ile the next day.

Those along the shore who witnessed the storm that night said that it possessed a fury never before seen on the lake. Some claimed that it pursued the vessel up the lake. By midnight, huge waves were battering the shores while the wind made banshee sounds as it tore through the trees.

Under such conditions, and considering the dignitaries aboard, it might have seemed natural for the captain to have sought shelter at some point; however, he steered single-mindedly toward Presqu'ile Bay. At the height of the storm, bonfires were lit along the shore to help guide the ship into safe harbor. Yet the captain paid them no heed. Nor did he appear to have control of his vessel, for her course seemed unerring. As if drawn by a huge magnet, the ship headed directly for the area of the monolith, then was lost from sight as the storm closed over the scene.

When the *Speedy* failed to make port next day, a search was begun along the shore. When the skies cleared, ships took up the search on the lake and passed the word to the American side that a vessel was missing and searchers there combed the beaches for wreckage.

As days passed without any news of the ship or her passengers, it was decided to drag the area around the monolith in case the ship had collided with the dangerous obstruction and gone to the bottom. But the searchers were in for a surprise. When they reached the location off Presqu'ile Bay they were unable to find any trace of the monolith. When they dragged the bottom they were even more surprised because there were no three-hundred-foot depths in the area. The entire thing had been filled in overnight! The bottom was now shallow and sandy.

The loss of the *Speedy* decimated the government and courts of Upper Canada, wiping out the cream of officialdom overnight. And it had done more. People were now beginning to question many of the strange elements in the whole affair. They wanted to know what the stone monolith had been; if it had been there over the years why had this single storm toppled it? Why had the *Speedy* headed directly for that area when her captain could have grounded her and saved all aboard? Further, what had filled in the three-hundred-foot hole so suddenly?

No one had the answers.

By the law of averages, there should have been a very large pile of stones on the bottom if the *Speedy* had collided with the monolith and toppled it. There was no such pile. If the ship had hit the monolith, then it should have sunk on or near the site and the debris and wreckage should have turned up on the nearby shore. There was no such wreckage that could be linked to this ship.

If neither the monolith nor the ship was in the area, then the natural question is: What kind of force could transport the entire multi-ton stone slab and the vessel to regions unknown?

Such a force is beyond what we normally think of as being natural; it is supernatural!

Considering all the various elements that were combined in this single event, it is a strange coincidence that would gather them all in one place. Ogetonicut never died the death of a murderer. Those who sought to impress their ideas on the young government of Upper Canada were removed, and the development of the north side of the lake was set back years by the disaster.

It seems as if the monolith acted like a magnet. When it had gathered all the elements together it eliminated them, then it was no longer needed and it, too, was eliminated.

It will never be known for sure whether the *Speedy* was still under the control of Captain Paxton as she was swept to the site of the monolith, or whether she was under the control of some unknown outside force. If it was the latter, then she isn't the only ship to find herself possessed by masters not of this world. For some, there was the attraction that drew them onto rocky shores; others simply sailed into a fog and vanished; still others were reduced to kindling.[20]

How sad for history that the monolith near Presqu'ile Bay seems to have been toppled over. Had the ship hit the monolith and made it fall to the bottom of the lake? Had the ship sunk with it and the storm filled up the bay with layers of mud?

Or perhaps the obelisk had been some interdimensional structure that had disappeared into the mists of time after the storm had passed.

Perhaps the remains of the obelisk are lying on the cold floor of the lake, waiting for a storm to move the mud and muck, and an underwater drone will come passing by to bathe it in the light of a flood lamp and take some photos for the local newspaper.

If we are to take this story at face value, and I do, then we have to wonder who placed this obelisk on what would become the lakebed of a giant body of water and when? Was the Presqu'ile Bay obelisk part of some now-lost worldwide network of obelisks and standing stones? Were these monolithic granite towers some kind of antennas used to broadcast and receive waves of energy? Perhaps it was a similar system as envisioned by the great inventor Nikola Tesla. Let us now look at Tesla, some of his inventions, and the feasibility of a worldwide power system utilizing obelisks as power towers.

According to the book *Obelisk: A History* a small plaque was discovered in Iowa in the late 1800s depicting animals along with two obelisks. Evidence of obelisks in the Americas?

The Washington Monument being struck by lightning.

An illustration from a Masonic book of the 1800s showing an obelisk at Alexandria in Egypt with Masonic symbols on stone blocks.

Nikola Tesla's experimental tower in Colorado Springs, 1899.

Chapter 8

The Towers of Atlantis

Arthur C. Clarke's Law of Revolutionary Ideas:
Every revolutionary idea—in science, politics, art, or
whatever—
evokes three stages of reaction in a hearer:
1) It is completely impossible—don't waste my time.
2) It is possible, but it is not worth doing.
3) I said it was a good idea all along.

I theorize in my book *Atlantis and the Power System of the Gods*[18] that a sophisticated civilization with electricity, power tools, machines and aircraft might have used obelisks in a wireless power system similar to that dreamed of by the famous inventor Nikola Tesla. Starting around 1918 Tesla began to promote his wireless power system that would radiate from towers that he had designed. These towers, hooked up to Tesla's 3-phase rotating magnetic field AC system, would broadcast usable power at a special wavelength that ships, airships, cars and trains could obtain and use to run their own electric motors.

The Terrible Crystals of Atlantis

The subject of Atlantis and a worldwide power system leads us to look for esoteric sources on these "power towers of Atlantis." These obelisks of power are first mentioned in the book *A Dweller on Two Planets*[41] and then later referred to as the "terrible crystal" in the psychic readings done in Virginia Beach by Edgar Cayce. As we shall see shortly, it is not entirely clear what these "terrible crystals"

265

were, but they seem to be some sort of gigantic vibrating crystal tower which seems to be a granite obelisk that is infused with tiny quartz crystals.

According to books published on Atlantis in the late 1800s these people possessed advanced technology that included electricity, hard metals, sonics and sound machines and even aircraft. They also possessed towers or obelisks that put some sort of energy into the atmosphere. These gigantic crystal towers, or terrible crystals, seem to have been similar to the system of broadcasting electricity from towers as proposed by the noted inventor Nikola Tesla.

The knowledge of these "terrible crystals" and their strange technology comes largely from metaphysical texts such as those of Edgar Cayce, the Theosophical Society, the Lemurian Fellowship and other similar groups. In this book we will focus largely on three sources, the teachings of Edgar Cayce and the books *A Dweller on Two Planets* and *An Earth Dweller Returns*.

The Incredible Maxt Tower of Atlantis

The power-towers of Atlantis, and the airships that drew power from them, were featured in great detail in two unusual books, *A Dweller on Two Planets*[41] and *An Earth Dweller Returns*.[42]

A Dweller on Two Planets was supposedly first dictated in 1884 by "Phylos the Thibetan" to a young Californian named Frederick Spencer Oliver who wrote the dictations down in manuscript form in 1886. The manuscript was not published until 1899, when it was finally released as a book. In 1940, the sequel, *An Earth Dweller Returns*, was published by The Lemurian Fellowship of Ramona, California. Also accredited to "Phylos the Thibetan" this book was allegedly dictated to Beth Nimrai. Both books are the long and complicated history of a number of persons and the karma created by each of them during their many lives, especially the karmic relationships and events of the "amanuensis" Frederick Spencer Oliver and his different lives as Rexdahl, Aisa and Mainin with the many lives of "Phylos" as Ouardl, Zo Lahm, Zailm and Walter Pierson, who was shown a secret door on the slopes of Mount Shasta, a mysterious mountain in northern California. Much of the book takes place in the Shasta region while other parts take place in Atlantis.

A map of Atlantis from the book *An Earth Dweller Returns.*

Both books are often difficult to follow, describing the past lives and the cycles of karma and rebirth between no less than eight people, men and women, including Beth Nimrai, the amanuensis of the later book *An Earth Dweller Returns.* This book is largely an attempt to correct and clarify much of the material in *A Dweller on Two Planets* and both books contain a great deal of detailed information on the life, times, culture and technology of ancient Atlantis, including the airships which were called vailxi in plural and vailx in singular.

A Dweller on Two Planets has remained a popular occult book for nearly a century largely because it contains detailed descriptions of devices and technology that were unquestionably well in advance of the time frame in which it was written. As the cover of one of the editions of the book states, "One of the greatest wonders of our times is the uncanny way in which *A Dweller on Two Planets* predicted inventions which modern technology fulfilled after the writing of the book."[41]

Among the inventions and devices mentioned in both books are air conditioners, to overcome deadly and noxious vapors; airless cylinder lamps, tubes of crystal illuminated by the "night side forces"; electric rifles, guns employing electricity as a propulsive force (rail-guns are a similar, and very new invention); monorail

267

transportation; water generators, instruments for condensing water from the atmosphere; and the vailx, an aerial ship governed by forces of levitation and repulsion.

Many of the words and terms in the books are identical to those in later Edgar Cayce readings, such as "night side forces" and the term

The Maxt Tower of Atlantis from *A Dweller on Two Planets*.

"Poseid" for Atlantis. While verification of any of the information in either book is impossible, the material is fascinating and of interest to any student of ancient sciences. In chapter two of *A Dweller on Two Planets,* the hero, Zailm (an earlier incarnation of Phylos and Walter Pierson), visits Caiphul, the capital of Atlantis, and views many wonderful electronic devices and the monorail system.

In chapter four, the electromagnetic airships of Atlantis are introduced along with radio and television (don't forget, this book was written in 1886). It is explained that the airships, similar to zeppelins, but more like a cigar-shaped airship, are electro-magnetic-gravitational and are capable of entering the water as a submarine or traveling through the air. Later, in chapter sixteen, Zailm takes a journey via vailx to "Suern" which is apparently ancient India or thereabouts.

In chapter eighteen Zailm visits the "Umaurean" (present day America) colonies of Poseid. In a fascinating portion of the book, the vailx stops for the night to visit a building on the summit of the Tetons. According to the text, "On the tallest of these had stood, perhaps for five centuries, a building made of heavy slabs of granite. It had originally been erected for the double purpose of worship of Incal (the Sun, or God), and astronomical calculations, but was used in my day as a monastery. There was no path up the peak, and the sole means of access was by vailx."

Frederick Spencer Oliver then alleges in a break in the story that such massive, granite slab-walls were discovered in 1886 by a Professor Hayden, allegedly the first person to climb Grand Teton. Whether such massive granite slabs, certainly in poor condition and probably thought to be naturally occurring, do indeed exist on or near the summit of Grand Teton, I have no way of knowing.

Afterward they visit the ancient copper mines of the Lake Superior region (which do indeed exist and are archaeological fact, though not satisfactorily explained) and then return to Poseid, making part of the journey underwater.

Back in Atlantis (Poseid) Zailm makes the mistake of getting involved with two women at the same time and karmic repercussions are severe when, about to marry one of the women, the other exposes him and tragedy follows when both women are killed. One commits suicide and the other stands in the Maxin Light, a kind of super energy beam in the center of the great temple, analogous to that mentioned in the similar Edgar Cayce reading (440-5; Dec. 20, 1933).

This Maxin Light apparently had something to do with the giant energy towers and "terrible crystals" of Atlantis. Perhaps it was an giant arc of power coming off one of the towers. These towers would also have functioned in a manner similar to that designed by the great inventor Tesla.

In chapter eighteen, Zailm speeds away in his private vailx and wanders for a time searching for gold in South America, using an electronic mineral detector, a water generator and an electric rifle. While searching for gold he is trapped in a small cavern by the evil priest Mainin (who is an early incarnation of Frederick Spencer Oliver, the amanuensis of the book) and dies.

A few incarnations later, Zailm (Phylos) is taken astral traveling to Venus, hence the title of the book, *A Dweller on Two Planets*. Though a small section, this part of the book is somewhat reminiscent of the Hari Krishna publication *Easy Journey To Other Planets* by Swami Prabhupada.[47]

In the second book, *An Earth Dweller's Return*,[42] much of the text is used to explain elements of *A Dweller on Two Planets* that had been left unexplained, particularly the karmic relationship between Phylos himself and Frederick Spencer Oliver. However a great deal of this book goes into the science of Atlantis including the cause of gravitational attraction; heat, magnetism and motion; transmutation of matter; the Maxin Light; a huge obelisk energy tower known as the Maxt; an airless cylinder light; levitation; and much more.

Part Five of the book is entitled "Description Of A Journey By vailx." According to the text, "the Atlantean vailx was an air vessel motivated by currents derived from the Night Side of Nature." Says the book:

> Altitude was dependent wholly upon pleasure. For this reason wide views were possible with a great variety of scenery. The rooms of the vailx were warmed by *Navaz* (Night Side of Nature) forces and furnished with the proper density of air by the same means. So rapidly did the aspect of things change beneath, that the spectator, looking

270

backwards, gazed upon a dissolving view.

The currents, derived from the Night Sides of Nature, permitted the attainment of the same rate of speed as the diurnal rotation of the Earth. For example, suppose we were at an altitude of ten miles and that the time was the instant of the sun's meridian. At that meridian moment, we could remain indefinitely bows on, while the Earth revolved beneath at approximately seventeen miles per minute. Or the reverse direction keys could be set, and our vailx would rush away from its position at the same almost frightful speed—

The Maxt Tower of Atlantis from *An Earth Dweller Returns.*

frightful to one unused to it, but not so to the returning Atlanteans who, in the Aquarian Age to come, will travel the highways of the land, sea and air without a thought of fear.[42]

During the trip, the vailx is beset by a storm and the night side forces are used. They sound a lot like a charged craft that moves through the water or the air by an electrically charged anti-gravity:

> The repulse keys were set, and presently we were so high in the air that all about our now closed ship were cirrus clouds—clouds of hail held aloft by the uprising of the winds which were severe enough to have been dangerous had our vessel been propelled by wings, fans, or gas reservoirs.
>
> But as we derived our forces of propulsion and repulsion from Nature's Night Side, or in Poseid phraseology, from *Navaz*, our long white aerial spindles feared no storm however severe. …The evening had not far advanced when it was suggested that the storm would most likely be wilder near the Earth, and so the repulse keys were set to a fixed degree, making nearer approach to the ground impossible as an accidental occurrence.

The chapter continues to speak of the journey, mentioning the destination, Suernis (India) and the air dispensers with wheels and

A vailx descends into the water from *A Dweller on Two Planets*.

pistons that pressurize the cabin. In Section 418 of the text it states:

The vailx used was about the middle traffic size. These vessels were made in four standard lengths: number one, about twenty-five feet; number two, eighty feet; number three about one hundred fifty-five feet; while the largest was approximately three hundred feet in length.

These long spindles were round, hollow needles of aluminum, comprising an outer and an inner shell between which were placed many thousands of double 'T' braces, an arrangement productive of intense rigidity and strength. Other partitions made other braces of additional resistant force. From amidships the vessels tapered toward either end to sharp points. Most vailxi were provided with an arrangement which allowed an open promenade deck at one end. The vailx which Zailm used was about fifteen feet and seven inches in diameter.

Crystal windows of enormous resistant strength were arranged in rows like port holes along the sides, with a few on top and several others set in the floor, thus affording a view in all directions.[42]

What is fascinating in reading descriptions of so-called Atlantean vailxi in these books, as well as their brief descriptions in the Edgar Cayce readings, is their similarity to the descriptions of vimanas in

An electric carriage in Atlantis as depicted in *A Dweller on Two Planets.*

273

ancient Indian texts and to a number of UFO craft seen in present times, including the late 1800s.

As the back cover paragraph on one edition of *A Dweller on Two Planets* points out, this book, and its description of a long, cylindrical, cigar-shaped aircraft, is a haunting premonition of not only many UFOs seen today, but of a type of craft that may yet be produced by a manufacturer in the near future!

That some of the UFO sightings of similar type craft of the past forty years might somehow be Atlantean or Indian-type vailx or vimana is a fantastic notion that apparently has never been considered by either the scientific community or by current UFO investigators.

The Great Crystal of Edgar Cayce

Similar information to that in the Phylos books comes from the "psychic" information reported from Edgar Cayce and the Association for Research and Enlightenment (A.R.E.) in Virginia Beach, Virginia. This association was founded as a library and center to study Edgar Cayce's many dictations about past lives, the sciences of Atlantis and Egypt, and alternative health information.

Known as the "sleeping clairvoyant," Edgar Cayce was born on March 18, 1877, on a farm near Hopkinsville, Kentucky. Even as a child he displayed powers of perception that seem to extend beyond the normal range. In 1898 at the age of twenty-one he became a salesman for a wholesale stationery company and developed a gradual paralysis of the throat muscles which threatened the loss of his voice. When doctors were unable to find a cause for the strange paralysis, he began to see a hypnotist. During a trance, the first of many, Cayce recommended medication and manipulative therapy that successfully restored his voice and cured his throat trouble.

He began doing readings for people, mostly of a medical nature, and on October 9, 1910, *The New York Times* carried two pages of headlines and pictures on the Cayce phenomenon. By the time Edgar Cayce died on January 3, 1945, in Virginia Beach, Virginia, he left well over 14,000 documented stenographic records of the telepathic-clairvoyant statements he had given for more than 8,000 different people over a period of 43 years. These typewritten documents are referred to as "readings." Important to our discussion in this book is

274

Edgar Cayce when he was younger.

that many of these "readings" concern Atlantis, persons' former lives in Atlantis, and the airships and crystals used in Atlantis. He also speaks of "nightside" forces, here spelled as one word.[65]

In reading 2437-1; Jan. 23, 1941, Cayce told his subject:

> ...[I]n Atlantean land during those periods of greater expansion as to ways, means and manners of applying greater conveniences for the people of the land—things of transportation, the aeroplane as called today, but then as ships of the air, for they sailed not only in the air but in other elements also. [65]

A number of persons who came to Cayce for individual life readings were, according to Cayce's readings, once navigators or engineers on these aircraft. He referred to the subject of the reading as "entity":

> [I]n Atlantean land when there were the developments of those things as made for motivative forces as carried the peoples into the various portions of the land and to other lands. Entity a navigator of note then. (2124-3; Oct. 2, 1931)
>
> ...[I]n Atlantean land when peoples understood the law of universal forces entity able to carry messages through space to the other lands, guided crafts of that period. (2494-1; Feb. 26, 1930) [65]

Cayce called the motive power used in these vessels the "nightside of life."

> [I]n Atlantean land or Poseidia—entity ruled in pomp and power and in understanding of the mysteries of the application of that often termed the nightside of life, or in applying the universal forces as understood in that period. (2897-1; Dec. 15, 1929)
>
> ...[I]n Atlantean period of those peoples that gained much in understanding of mechanical laws and application of nightside of life for destruction. (2896-1; May 2, 1930) [65]

Cayce speaks of the use of crystals or "firestones" for energy and related applications. He also speaks of the misuse of power and warnings of destruction to come:

> [I]n Atlantean land during the periods of exodus due to foretelling or foreordination of activities which were bringing about destructive forces. Among those who were not only in Yucatan but in the Pyrenees and Egyptian land, for the manners of transportation and communications through airships of that period were such as Ezekiel described at a much later date. (4353-4; Nov. 26, 1939. See *Ezekiel* 1:15-25, 10:9-17 RSV.)
>
> ...[I]n Atlantis when there were activities that brought about the second upheaval in the land. Entity was what would be in the present the electrical engineer—applied those forces or influences for airplanes, ships, and what you would today call radio for constructive or destructive purposes. (1574-1; April 19, 1938)
>
> ...[I]n Atlantean land before the second destruction when there was the dividing of islands, when the temptations were begun in activities of Sons of Belial and children of the Law of One. Entity among those that interpreted the messages received through the crystals and the fires that were to be the eternal fires of nature. New developments in air and water travel are no surprise to this entity as these were beginning development at that period for escape. (3004-1; May 15,

1943)

...[I]n Atlantean land at time of development of electrical forces that dealt with transportation of craft from place to place, photographing at a distance, overcoming gravity itself, preparation of the crystal, the terrible mighty crystal; much of this brought destruction. (519-1; Feb. 20, 1934)

...[I]n city of Peos in Atlantis—among people who gained understanding of application of nightside of life or negative influences in the Earth's spheres, of those who gave much understanding to the manner of sound, voice and picture and such to peoples of that period. (2856-1; June 7, 1930)

...[I]n Poseidia the entity dwelt among those that had charge of the storage of the motivative forces from the great crystals that so condensed the lights, the forms of the activities, as to guide the ships in the sea and in the air and in conveniences of the body as television and recording voice. (813-1; Feb. 5, 1935)[65]

The use of crystals as an important part of the technology is mentioned in a long reading from Dec. 29, 1933:

About the firestone—the entity's activities then made such applications as dealt both with the constructive as well as destructive forces in that period. It would be well that there be given something of a description of this so that it may be understood better by the entity in the present.

In the center of a building which would today be said to be lined with nonconductive stone—something akin to asbestos, with ...other nonconductors such as are now being manufactured in England under a name which is well known to many of those who deal in such things.

The building above the stone was oval; or a dome wherein there could be ...a portion for rolling back, so that the activity of the stars—the concentration of energies that emanate from bodies that are on fire themselves, along with elements that are found and not found in the Earth's atmosphere.

The concentration through the prisms or glass (as would be called in the present) was in such manner that it acted upon the instruments which were connected with the various modes of travel through induction methods which made much the [same] character of control as would in the present day be termed remote control through radio vibrations or directions; though the kind of force impelled from the stone acted upon the motivation forces in the crafts themselves.

The building was constructed so that when the dome was rolled back there might be little or no hindrance in the direct application of power to various crafts that were to be impelled through space—whether within the radius of vision or whether directed under water or under other elements, or through other elements.

The preparation of this stone was solely in the hands of the initiates at the time; and the entity was among those who directed the influences of radiation which arose, in the form of rays that were invisible to the eye but acted upon the stones themselves as set in the motivating forces—whether the aircraft were lifted by the gases of the period; or whether for guiding the more-of-pleasure vehicles that might pass along close to the Earth, or crafts on the water or under the water.

These, then, were impelled by the concentration of rays from the stone which was centered in the middle of the power station, or powerhouse (as would be the term in the present).

In the active forces of these, the entity brought destructive forces by setting up—in various portions of the land—the kind that was to act in producing powers for the various forms of the people's activities in the same cities, the towns, and the countries surrounding same. These, not intentionally, were tuned too high; and brought the second period of destructive forces to the people of the land—and broke up the land into those isles which later became the scene of further destructive forces in the land.

Through the same form of fire the bodies of individuals

were regenerated; by burning—through application of rays from the stone—the influences that brought destructive forces to an animal organism. Hence the body often rejuvenated itself; and it remained in that land until the eventual destruction; joining with the peoples who made for the breaking up of the land—or joining with Belial, at the final destruction of the land. In this, the entity lost. At first it was not the intention nor desire for destructive forces. Later it was for ascension of power itself.

As for a description of the manner of construction of the stone: we find it was a large cylindrical glass (as would be termed today); cut with facets in such manner that the capstone on top of it made for centralizing the power or force that concentrated between the end of the cylinder and the capstone itself. As indicated, the records as to ways of constructing same are in three places in the Earth, as it stands today: in the sunken portion of Atlantis, or Poseidia, where a portion of the temples may yet be discovered under the slime of ages of sea water—near what is known as Bimini, off the coast of Florida. And (secondly) in the temple records that were in Egypt, where the entity acted later in cooperation with others towards preserving the records that came from the land where these had been kept. Also (thirdly) in records that were carried to what is now Yucatan, in America, where these stones (which they know so little about) are now—during the last few months—being uncovered. (440-5; Dec. 20, 1933)[65]

What Cayce was apparently describing was a Tesla-type system using an obelisk with a capstone to transmit power to the world—the terrible crystal. This would basically be identical to the Maxt Tower described in the Phylos books. These towers would have been in a number of places, each broadcasting power at a special

An older Edgar Cayce.

resonance to the ships and airships that were in motion and drawing power from this resonant grid.

Perhaps Cayce's terrible crystal was one single source such as the Great Pyramid of Egypt. That pyramid has a unique configuration that may have made it a gigantic maser capable of powering satellites and cities. Other pyramids in the Giza and Saqqara areas, including the Red Pyramid, may have also been chemical masers as engineer and author Christopher Dunn theorizes was in his books, to be discussed shortly. But first let us look at Tesla's power system and electric resonance.

Electric Resonance for Power Generation

While the generation of AC current is really the result of rotating magnetic fields, the theory of transmitting power through the atmosphere in a similar manner as radio waves was a pet project for the great inventor Nikola Telsa. A contemporary of Tesla's was John Worrell Keely, an inventor from Philadelphia who claimed that resonance was the key to all sorts of things, including power generation and transmission, anti-gravity and free energy. Keeley (1837-1898) was the inventor of such mysterious devices as the "Vibrodyne Motor," the "Compound Disintegrator," and the "Provisional Engine." Much of his work was kept a secret, though he made frequent demonstrations to eager investors. Not much happened in the end, and Keeley has been branded a fraud-artist.

We may never know the validity of Keeley's inventions, but Tesla—the world's greatest inventor—believed that resonance was key to power generation and transmission. Tesla wanted to take power that was being generated from the Niagara Falls Hydroelectric Station and broadcast the power from towers that he was building on Long Island. In theory, Tesla could have used obelisks to broadcast his power. If so, resonance may have been key to the obelisk functioning properly.

The following article was printed in the Indian newspaper *The Hindu, Science & Technology Supplement,* on November 20, 1997. It is a translation of a Russian article written by Konstantin Smirnov for the magazine *RIA Novosti*:

When Dr. Andrei Melnichenco, a physicist specializing

in electrodynamics in the city of Chekhov near Moscow, called our editorial office and described his invention, I did not believe him. But my mistrust did not perplex the inventor, and he offered to demonstrate his device.

The device consists of several batteries and a small converter to change direct current into alternating current (220V, 50Hz) using an electric motor.

The power of this motor is far greater than that of the power source. When a small plate with several assemblies is added to the chain of components and switched on, the motor begins to pick up speed in such a way that it would be possible to set an abrasive circle on it and sharpen a knife.

In another experiment, a fan serves as the final component of the device. At first, its blades are slowly rotating but, after a special unit is connected in sequence with it, the fan immediately gains speed and makes a good 'breeze'. All this looked strange, primarily from the standpoint of the law of conservation of energy.

Seeing my perplexity, Melnichenko explained that the processes taking place in his device are simple enough, and are based on the phenomenon of electric resonance.

Despite the fact that this phenomenon has been known for more than a century, it is only rarely used in radio engineering and communications electronics where amplification of a signal by many times is needed.

Resonance is not used much in electrical engineering and power generation. By the end of the last century, the great scientist Nicola Tesla used to say that without resonance, electrical engineering was just a waste of energy.

No one attached any importance to this pronouncement at that time. Many of Tesla's works and experiments, for instance the transmission of electricity by one unearthed wire, have only recently been explained.

The scientist staged these experiments a century ago, but it has only been in our days that S. V. Avramenko has managed to reproduce them. This also holds true for the transmission of electric power by means of electromagnetic waves and resonance transformers.

281

"My first experiments with high-frequency resonance transformers produced results which, to say the least, do not always accord with the law of the conservation of energy, but there is a simple mathematical and physical explanation of this," Melnichenko says.

"I have designed several special devices and electric motors which contain many of these ideas and which may help them achieve full resonance in a chain when it consumes energy only in the form of the thermal losses in the winding of the motor and wires of the circuits while the motor rotates without any consumption of energy whatsoever.

"This was shown during the demonstration," the inventor goes on to say. "The power, supplied to the motors, was less than was necessary for their normal operation! I have called the new physical effect transgeneration of electric power. Electric resonance is the principle underlying the operation of the device."

This effect can be very widely used. For instance, electric resonance motors may be employed in electric cars. In this case the storage batteries' mass is minimal.

The capacity, developed by an electric motor, exceeds the supplied electric power by many times, which may be used for devising absolutely autonomous propulsion power units—a kind of superpower plant under the hood.

The battery-driven vehicles, equipped with such power plants, would not need frequent recharging because, just as in the case of an ordinary engine, it would only need storage batteries for an electric start. All the results have been confirmed by hundreds of experiments with resonances in electric motors (both ordinary and special).

In special motors, it is possible to achieve the quality of resonance in excess of 10 units. The technology of their manufacture is extremely simple while the investments are minimal. The results are superb!

Electromechanics is only the first step. The next are statical devices, which are resonance-based electric power generators. For instance, a device, supplied at the input with power equal to that of three 'Energizer' batteries can make a

100-watt incandescent lamp burn at the exit.

The frequency is about 1 MHz. Such a device has a rather simple circuit, and is based on resonance. Using it, it is possible to by far increase the power factor of energy networks, and to drastically cut the input (reactive) resistance of ordinary transformers and electric motors.

But creation of fundamentally new, environmentally clean electric power generators is the most important application of electric resonance.

A resonance-based energy transformer will become the main element of such devices. The employment of conductors with very low active resistance—cryoelectrics—for their windings will make it possible to increase power by hundreds and thousands of times, in proportion to resonance qualities of the device.

The Russian Academy of Sciences, in its review says that the principle underlying the operation of the devices does not rouse doubts in theory and in practice, and that the work of the resonance-based electric systems is not in conflict with the laws of electrophysics.

According to the above article, resonance is key to proper power generation, transmission and reception. If crystalline granite towers were used for a Tesla-type system they would need to be "tuned" like giant tuning forks. They would need to be "tuned" to the resonance of the electric power. Obelisks made of solid granite could fulfill this function.

According to the San Diego spiritual group Unarius—believers in reincarnation and Atlantis and having a fascination with Tesla—Nikola Tesla was the reincarnation of an Atlantean engineer and inventor who was responsible for the energy supply first used to provide power on a now destroyed island in the Atlantic. In Unarius theory, from the great central pyramid in Atlantis, power beams would be relayed from reflectors on mountaintops into the different homes where these power beams would be converted into light, heat or even used to cool the house. They said that every house would have a round glass globe or sphere about a foot in diameter that was filled with certain rare gases that would fluoresce and give off a soft white light, just as does a modern fluorescent light. Heating or cooling was also quite simple: Air being made up of molecules of gases, electrical energy of a certain frequency could be radiated through the air and converted into heat through "hysteresis" in the electromagnetic fields of the atoms.

There is a famous photograph of Nikola Tesla holding up a light which is lit but does not have any wires connecting it to a power

A portrait of Nikola Tesla from 1919.

source. Instead, it is drawing power directly from a transmitting tower nearby. Perhaps in ancient times such simple lights were in many dwellings and at night they glowed a dull yellow or a dull green and gave light during the daily hours of darkness much as we now enjoy because of power lines strung across the land on high towers.

The same proposition in reverse makes the air become cold. Similarly, the atmosphere on the Earth is always converting certain electromagnetic energy into heat. Speaking from the point of absolute zero (-460 degrees Fahrenheit), all air on the surface of the Earth is comparatively warm, even at the poles.

Cooling or heating the air at any given point means merely to decrease or increase the "electromagnetic hysteresis." As a definition for a Pabst hysteresis-synchronous motor, Unarius says that it is the "inductive principle of cosmic hysteresis," and adds that "The reference to 'hysteresis' is not the earth-electronics definition, but rather an electromagnetic conversion process wherein cyclic (4th dim.) waveform-structures are transformed into lower (3rd dim.) waveform-structures."

Tesla and the Power System of Atlantis

The Serbian-American electrical engineer and inventor Nikola Tesla was born in Smijlan, Croatia (then part of Austria-Hungary) on July 9, 1856, the son of a clergyman and an inventive mother. It was Tesla who devised the alternating-current systems that underlies the modern electrical power industry.

Tesla had an extraordinary memory, one that made learning six languages easy for him. He entered the Polytechnic School at Gratz, where for four years he studied mathematics, physics and mechanics, confounding more than one professor by an understanding of electricity, an infant science in those days, that was greater than theirs. His practical career started in 1881 in Budapest, Hungary, where he made his first electrical invention, a telephone repeater (the ordinary loudspeaker), and conceived the idea of a rotating magnetic field, which later made him world famous in its form as the modern induction motor. The polyphase induction motor is what provides power to virtually every industrial application, from conveyer belts to winches to machine tools.

Tesla immigrated to the USA in 1884, bringing with him the various models of the first induction motors. He worked briefly for Thomas Alva Edison, who as the advocate of direct current became Tesla's unsuccessful rival in electric-power development. In 1888, Tesla showed how a magnetic field could be made to rotate if two coils at right angles were supplied with alternating currents 90 degrees out of phase with each other. Tesla's induction motors were eventually shown to George Westinghouse. Westinghouse bought rights to the patents on this motor and made it the basis for the Westinghouse power system. It was in the Westinghouse shops that the induction motor was perfected. Numerous patents were taken out on this prime invention, all under Tesla's name.

During his short tenure working under Thomas Edison, Tesla created many improvements on Edison's DC motors and generators, but left under a cloud of controversy after Edison refused to live up to bonus and royalty commitments. This was the beginning of a rivalry which was to have ugly consequences later when Edison and his backers did everything in their power to stop the development and installation of Tesla's far more efficient and practical AC current delivery system and urban power grid. Edison put together a traveling road show which attempted to portray AC current as dangerous, even to the point of electrocuting animals both small (puppies) and large (in one case an elephant) in front of large audiences. As a result of this propaganda crusade, the state of New York adopted AC electrocution as its method of executing convicts. Tesla won the battle by the demonstration of AC current's safety and usefulness when his apparatus illuminated and powered the entire New York World's Fair of 1899.

It was Tesla's association and loose partnership with George Westinghouse that brought about the implementation of Tesla's amazing inventions, including the alternating current power we use today.

Westinghouse was born in Central Bridge, New York on Oct. 6, 1846, and was an American inventor and industrialist who during his lifetime obtained approximately 400 patents, including that on the air brake. In 1865 he patented a device for replacing derailed freight cars on tracks; three years later he developed the railroad frog, which permits the wheel on one rail of a track to cross another rail

286

The Tesla Company letterhead featuring his tower for wireless power transmission.

of an intersecting track. In 1869 he founded the Westinghouse Air Brake Company. By the early 1880s, Westinghouse had developed interlocking switches and a complete railroad signal system and established (1882) the Union Switch and Signal Company. In 1886 he founded the Westinghouse Electric Company, which established the use of alternating current for electrical generating and transmitting apparatuses and electrical appliances in the United States. Westinghouse died on March 12, 1914.

Without the guidance and business sense of Westinghouse, Tesla became increasingly ineffective in his ability to get his inventions funded and released to the public. Tesla's other inventions included the Tesla coil, a kind of transformer, and he did notable research on high-voltage electricity and wireless communication. He made little money from his work, however, and later lived as an eccentric recluse.

Tesla died in New York City on January 7, 1943, many of his dreams unrealized. On his death the FBI searched Tesla's apartment and seized various papers and technical diagrams that he had kept. What inventions were seized by the FBI is not currently known.

Electric power has become an indispensable form of energy throughout much of the world. Even systems that use forms of energy other than electricity are likely to contain controls or equipment that run on electric power. For example, modern home heating systems may burn natural gas, oil, or coal, but most systems have combustion and temperature controls that require electricity in order to operate. Similarly, most industrial and manufacturing

287

processes require electric power, and the computers and business machines of many offices and commercial establishments are paralyzed if electric service is interrupted.

During the first part of the 20th century, only about 10% of the total energy generated in the United States was converted to electricity. By 1990 electric power accounted for about 40% of the total. Developing countries are usually not as dependent on electricity as are the more industrialized nations, but the growth rate of electricity use in some of those countries is comparable to the rate of growth in the early years of electricity availability in the United States.

The first commercial electric-power installations in the United States were constructed in the latter part of the 19th century. The Rochester, N.Y., Electric Light Co. was established in 1880. In 1882, Thomas A. Edison's Pearl Street steam-electric station began

operation in New York City and within a year was reported to have had 500 customers for the lighting services it supplied. A short time later a central station powered by a small waterwheel began operation in Appleton, Wisconsin.

In 1886 the feasibility of sending electric power greater distances from the point of generation by using alternating current (AC) was demonstrated at Great Barrington, Mass. The plant there utilized Westinghouse transformers to raise the voltage from the generators for a high-voltage transmission line.

The electric power industry of the United States grew from small beginnings such as these to become, in less than 100 years, the most heavily capitalized

Tesla's World Wireless logo.

industry in the country. It now comprises about 3,100 different corporate entities, including systems of private investors, federal and other government bodies, and cooperative-user groups. Less than one-third of the corporate groups have their own generating facilities; the others are directly involved only in the transmission and distribution of electric power.

Electric power transmission systems today consist of step-up transformer stations to connect the lower-voltage power-generating equipment to the higher-voltage transmission facilities; high-voltage transmission lines and cables for transferring power from one point to another and pooling generation resources; switching stations, which serve as junction points for different transmission circuits; and step-down transformer stations that connect the transmission circuits to lower-voltage distribution systems or other user facilities. In addition to the transformers, these transmission substations contain circuit breakers and associated connection devices to switch equipment into and out of service, lightning arresters to protect the equipment, and other appurtenances for particular applications of electricity. Highly developed control systems, including sensitive devices for rapid detection of abnormalities and quick disconnection of faulty equipment, are an essential part of every installation in order to provide protection and safety for both the electrical equipment and the public.

Many of the first high-voltage transmission lines in the United States were built principally to transmit electrical energy from

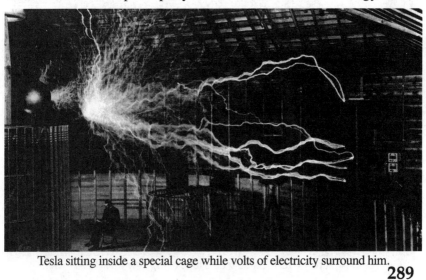

Tesla sitting inside a special cage while volts of electricity surround him.

289

hydroelectric plants to distant industrial locations and population centers. High-voltage transmission lines were originally designed to permit the construction of large generating units and central stations on attractive, remote sites close to fuel sources and supplies of cooling water. Today, however, they connect different power networks in order to achieve greater economy by exchanges of low-cost power, to achieve savings in reserve generating capacity, to improve the reliability of the system, and to take advantage of diversity in the peak loads of different systems and thereby reduce operating costs.

Before World War II the highest-voltage lines in the United States were 230 kV, with the exception of one 287-kV line from Boulder Dam to Los Angeles. In the early 1950s several 345-kV lines were constructed. By 1964 the first 500-kV lines in the United States were being completed, and in 1969 the first 765-kV line was put into service. All of these involved AC systems.

In 1970 a 1,380-km (856-mi), 800-kV direct-current (DC) line was placed in commercial service to connect northwestern U.S. hydroelectric sources with the Los Angeles area. Such systems offer an economical means of transferring large quantities of power

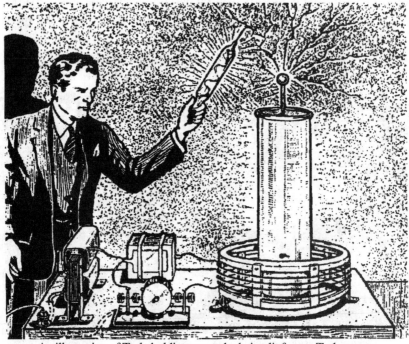

290   An illustration of Tesla holding up a tube being lit from a Tesla tower.

N. TESLA.
APPARATUS FOR TRANSMITTING ELECTRICAL ENERGY.
APPLICATION FILED JAN. 18, 1902. RENEWED MAY 4, 1907.

1,119,732.

Patented Dec. 1, 1914.

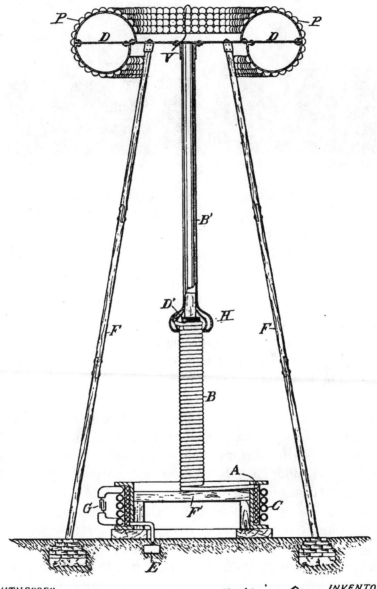

WITNESSES:

M. Lawson Dyer

Benjamin Miller.

INVENTOR,

Nikola Tesla,

BY Kerr, Page & Cooper

his   ATTORNEYS.

Tesla's 1914 patent for the wireless transmission of power.

The tower and power station partially built at Wardenclyffe, Long Island.

over long distances. They also avoid stability problems sometimes encountered by AC systems and DC systems are sometimes used to connect AC systems even over short transmission distances. Still, the transmission of power through wires has many limitations. Tesla decided that wireless transmission of power was the better way to go.

The Wireless Transmission of Power

Tesla's most important work at the end of the nineteenth century was his original system of transmission of energy by wireless antenna. In 1900 Tesla obtained his two fundamental patents on the transmission of true wireless energy covering both methods and apparatus and involving the use of four tuned circuits. In 1943, the Supreme Court of the United States granted full patent rights to Nikola Tesla for the invention of the radio, superseding and nullifying any prior claim by Marconi and others in regards to the "fundamental radio patent."

It is interesting to note that Tesla, in 1898, described the transmission of not only the human voice, but images as well and later designed and patented devices that evolved into the power supplies that operate our present day TV picture tubes. The first primitive radar installations in 1934 were built following principles—mainly regarding frequency and power level—that were stated by Tesla in 1917.

In 1889 Tesla constructed an experimental station in Colorado

Springs where he studied the characteristics of high frequency or radio frequency alternating currents. While there he developed a powerful radio transmitter of unique design and also a number of receivers "for individualizing and isolating the energy transmitted." He conducted experiments designed to establish the laws of radio propagation that are currently being "rediscovered" and verified amid some controversy in high energy quantum physics.

Tesla wrote in *Century Magazine* in 1900:

> ...that communication without wires to any point of the globe is practicable. My experiments showed that the air at the ordinary pressure became distinctly conducting, and this opened up the wonderful prospect of transmitting large amounts of electrical energy for industrial purposes to great distances without wires... its practical consummation would mean that energy would be available for the uses of man at any point of the globe. I can conceive of no technical advance which would tend to unite the various elements of humanity more effectively than this one, or of one which would more add to and more economize human energy...

Over a century ago Tesla had devised a system of broadcasting

An artist's concept of what Tesla's tower would have looked like in action.

electrical power through the atmosphere by utilizing a network of specially designed towers. This, I am arguing in this book, is essentially the same as the Atlantean power system in use many thousands of years ago.

After finishing preliminary testing, work was begun on a full-sized broadcasting station at Shoreham, Long Island. Had it gone into operation, it would have been able to provide usable amounts of electrical power at the receiving circuits. After construction of a generator building and a 180-foot broadcasting tower, financial support for the project was suddenly withdrawn by J. P. Morgan when it became apparent that such a worldwide power project couldn't be metered and charged for.

According to Toby Grotz of the (now defunct) International Tesla Society, it has been proven that electrical energy can be propagated around the world between the surface of the Earth and the ionosphere at extreme low frequencies in what is known as the Schumann Cavity. The Schumann Cavity surrounds the Earth at ground level and extends upward to a maximum 80 kilometers. Experiments to date have shown that electromagnetic waves of extreme low frequencies in the range of 8 Hz, the fundamental Schumann Resonance frequency, propagate with little attenuation around the planet within the Schumann Cavity. Knowing that a resonant cavity can be excited and that power can be delivered to that cavity similar to the methods used in microwave ovens for home use, it should be possible to resonate and deliver power via the Schumann Cavity to any point on Earth. This will result in practical wireless transmission of electrical power.

According to Grotz, although it was not until 1954-1959 when experimental measurements were made of the frequency that

A quartz crystal cut as a small obelisk.

294

is propagated in the resonant cavity surrounding the Earth, recent analysis shows that it was Nikola Tesla who, in 1899, first noticed the existence of stationary waves in the Schumann Cavity. Tesla's experimental measurements of the wave length and frequency involved closely match Schumann's theoretical calculations. Some of these observations were made in 1899 while Tesla was monitoring the electromagnetic radiations due to lightning discharges in a thunderstorm which passed over his Colorado Springs laboratory and then moved more than 200 miles eastward across the plains.

In his *Colorado Springs Notes,* Tesla noted that these stationary waves "...can be produced with an oscillator," and added in parentheses, "This is of immense importance." The importance of his observations is due to the support they lend to the prime objective of the Colorado Springs laboratory. The intent of the experiments and the laboratory Tesla had constructed was to prove that wireless transmission of electrical power was possible.

According to the International Tesla Society, Schumann Resonance is analogous to pushing a pendulum. A working Tesla wireless power system could create pulses or electrical disturbances that would travel in all directions around the Earth in the thin membrane of non-conductive air between the ground and the ionosphere. The pulses or waves would follow the surface of the Earth in all directions expanding outward to the maximum circumference of the Earth and contracting inward until meeting at a point opposite to that of the transmitter. This point is called the anti-pode. The traveling waves would be reflected back from the anti-pode to the transmitter to be reinforced and sent out again.

At the time of his measurements Tesla was experimenting with and researching methods for "...power transmission and transmission of intelligible messages to any point on the globe." Although Tesla was not able to commercially market a system to transmit power around the globe, modern scientific theory and mathematical calculations support his contention that the wireless propagation of electrical power is possible and a feasible alternative to the extensive and costly grid of electrical transmission lines used today for electrical power distribution.

This system would essentially require a network of power towers to pulse energies into the atmosphere. Generating stations would still be required to rotate the giant magnets that are pulsing out Tesla's three-phase AC power such as those at hydroelectric stations, thermal or other. One of these power plants may have been a massive maser.

The Giant Giza Maser

Before we depart from this subject were should look at the possibility that "the terrible crystal" was a single power plant that emitted power that the obelisks could use. The belief that the Great Pyramid at Giza was in fact a gigantic maser sending power to a satellite in geosynchronous orbit has been

Engineer and author Christopher Dunn.

championed by several authors, particularly by British engineer Christopher Dunn. Dunn is the author of the 1998 book *The Giza Power Plant: Technologies of Ancient Egypt*.[16] In this book, Dunn outlines his theories, and gives evidence for advanced machining and engineering knowledge in ancient Egypt. Another similar book is *The Giza Death Star* by Oklahoma physicist Joseph P. Farrell.[52]

Dunn claims that the Earth may be a giant power plant, and the many pyramids, obelisks, and megalithic standing stones may be part of this great "energy system." He says that the Great Pyramid was a giant power plant in the form of a maser and that harmonic resonators were housed in slots above the King's Chamber. He also theorized that there was a hydrogen explosion inside the King's Chamber that shut down the power plant's operation.

Says Dunn:

> While modern research into architectural acoustics might predominantly focus upon minimizing the reverberation effects of sound in enclosed spaces, there is reason to

believe that the ancient pyramid builders were attempting to achieve the opposite. The Grand Gallery, which is considered to be an architectural masterpiece, is an enclosed space in which resonators were installed in the slots along the ledge that runs the length of the Gallery. As the Earth's vibration flowed through the Great Pyramid, the resonators converted the energy to airborne sound. By design, the angles and surfaces of the walls and ceiling of the Grand Gallery, caused reflection of the sound and its focus into the King's Chamber. Although the King's Chamber was also responding to the energy flowing through the pyramid, much of the energy would flow past it. The design and utility of the Grand Gallery was to transfer the energy flowing through a large area of the pyramid into the resonant King's Chamber. This sound was then focused into the granite resonating cavity at sufficient amplitude to drive the granite ceiling beams to oscillation. These beams, in turn, compelled the beams above them to resonate in harmonic sympathy. Thus, the input of sound and the maximization of resonance, the entire granite complex, in effect, became a vibrating mass

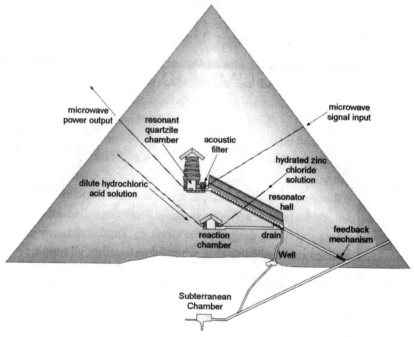

Dunn's concept of the Great Pyramid as a giant maser.

of energy.

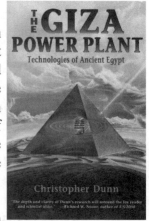

The acoustic qualities of the design of the upper chambers of the Great Pyramid have been referenced and confirmed by numerous visitors since the time of Napoleon, whose men discharged their pistols at the top of the Grand Gallery and noted that the explosion reverberated into the distance like rolling thunder.

Dunn says that it is possible to confirm that the Grand Gallery indeed reflected the work of an acoustical engineer using only its dimensions:

> The disappearance of the gallery resonators is easily explained, even though this structure was only accessible through a tortuously constricted shaft. The original design of the resonators will always be open to question; however, there is one device that performs in a manner that is necessary to respond sympathetically with vibrations. There is no reason that similar devices cannot be created today. There are many individuals who possess the necessary skills to recreate this equipment.[16]

According to Dunn, a Helmholtz resonator would respond to vibrations coming from within the Earth, and actually maximize the transfer of energy! The Helmholtz resonator is made of a round hollow sphere with a round opening that is 1/10 — 1/5 the diameter of the sphere. The size of the sphere determines the frequency at which it will resonate. If the resonant frequency of the resonator is in harmony with a vibrating source, such as a tuning fork, it will draw energy from the fork and resonate at greater amplitude than the fork will without its presence. It forces the fork to greater energy output than what is normal. Unless the energy of the fork is replenished, the fork will lose its energy quicker than it normally would without the Helmholtz resonator. But as long as the source continues to vibrate, the resonator will continue to draw energy from it at a greater rate.

298

Dunn says that the Helmholtz resonator is normally made out of metal, but can be made out of other materials. Holding these resonators in place inside the Gallery are members that are "keyed" into the structure by first being installed into the slots, and then held in the vertical position with "shot" pins that locate in the groove that runs the length of the Gallery.

Dunn now thinks that:

> The material for these members could have been wood, as trees are probably the most efficient responders to natural Earth sounds. There are trees that, by virtue of their internal structure, such as cavities, are known to emit sounds or hum. Modern concert halls are designed and built to interact with

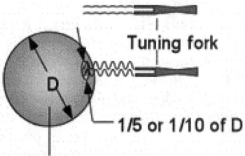

Tuning fork

1/5 or 1/10 of D

Helmholtz resonator

By virtue of its design, the Helmholtz resonator, over time, draws more energy from a vibrating source, such as a tuning fork, than what the source will give up naturally.

Photograph courtesy of Robert McKenty

Fine stonework in the Cairo Museum. Helmholtz resonators?

Dunn's explanation of a Helmholtz resonator.

299

the instruments performing within. They are huge musical instruments in themselves. The Great Pyramid can be seen as a huge musical instrument with each element designed to enhance the performance of the other.[16]

What Dunn is describing seems fantastic—a giant beam of maser energy was broadcast out of the pyramid to what he calls his "Eye of Horus" satellite which then distributed the power to other locations on the Earth—possibly to tuned crystal towers.

Essentially, the preceding stories offer us a curious "what if?"

What if there was an ancient power system—a power system that both gave power to, and helped guide, the airships of the time?

What if there was a large-scale system that utilized crystal broadcast towers to send power into the atmosphere?

We see real evidence of a worldwide power system that included obelisks and other towers. These monolithic quartz-infused towers were cut to a certain resonance and then placed at certain sites around the globe to receive and transmit energy. Were these ancient power towers on other nearby worlds like our Moon and Mars? Let us look into the astonishing evidence of obelisks and power towers off of planet Earth.

300   Dunn's concept of the Great Pyramid as a giant maser.

Chapter 9

Obelisks on the Moon

UFOs are astronautical craft, or entities.
If they have a fixed base of any kind, that base is likely the Moon.
—Morris Jessup

The Moon's composition is not at all what it should
be had the Moon been formed in its present orbit
around the earth.
-Dr. Harold Urey,
Science News, October 4, 1971

Obelisks and the Moon

Incredibly, there exists evidence for obelisks and pyramids on the Moon and possibly on the planet Mars and its moons. We also have the site of a monolith, discovered by the Soviets, that is similar to the mysterious rectangular block of stone that appears in *2001: A Space Odyssey*, Arthur Clarke's famous book, made into a movie by Stanley Kubrick.

While there may be a number of towers, pyramids and objects on the Moon, what specifically interests us is obelisks and our Moon seems to have a number of them, including a cluster of obelisks known as the Blair Cuspids. The Blair Cuspids are six formations that were captured in photographs taken by NASA's *Lunar Orbiter 2* in the Moon's Sea of Tranquility. The cuspids, or obelisks, have been compared in appearance to the Washington Monument. They are said to have precise shapes and patterns, and are in perfectly aligned geometric positions. They are named after anthropologist William Blair, who worked for the Boeing Institute of Biotechnology examining the photographs taken by *Lunar Orbiter 2*.

His comments about the cuspids appeared in a *Washington Post* article titled *Six Mysterious Statuesque Shadows Photographed on the Moon by Orbiter,* on November 23, 1966. In the article Blair stated, "If the cuspids really were the result of some geophysical event, it would be natural to expect to see them distributed at random. As a result, the triangulation would be scalene or irregular, whereas those concerning the lunar object lead to a basilary system, with coordinate x,y,z to the right angle, six isosceles triangles and two axes consisting of three points each."[49]

The article in the *Washington Post* on November 23, 1966, described the "mysterious statuesque shadows" as something like the Washington Monument:

> Six statuesque and mysterious shadows on the moon were photographed and relayed to earth yesterday by Lunar Orbiter 2.
>
> Ranging from one about 20 feet long to another as long as 75 feet, the six shadows were hailed by scientists as one of the most unusual features of the moon ever photographed.
>
> One scientist described the needle-like shadows as the moon's "Christmas tree effect." Still another description called it the "Fairy Castle" effect. On seeing the picture, one scientist wanted to call the region the moon's "Valley of Monuments."
>
> The region of the moon where the shadows turned up is just to the western edge of the moon's Sea of Tranquility. It is an area just north of the moon's equator, slightly to the east or right, of center.

A photograph of the Blair Cuspids.

Scientists said they have no idea what is casting the shadows. The largest shadow is just the sort that would be cast by something resembling the Washington Monument, while the smallest is the kind of shadow that might be cast by a Christmas tree.

Four of the shadows are clustered together and are on the slope of an "old" lunar crater—one that has been there for a long time, more than 1 million years.

All around the six shadows is the more familiar lunar landscape—the crater-marked face that gives the moon the appearance of a cooking pancake just before it is flipped over.

The picture was the most dramatic photo taken thus far by Orbiter 2 and came after the spacecraft transmitted two other "unusual" lunar views over the weekend.

One showed a large crater so close up that scientists said it was about three-fourths the size of the Rose Bowl. Another pictured a lunar rock field not unlike the sandlot field where youngsters might play baseball.

All three pictures were taken by Orbiter's 2-inch telephoto lens, which is capable of revealing objects on the moon the size of a manhole cover. The pictures were snapped from an altitude of about 30 miles or less.

A drawing of the Blair Cuspids from the photo.   **303**

Some months later, on February 1, 1967, the *Los Angeles Times* ran an article about the obelisks, citing the opinions of William Blair, an anthropologist and member of the Boeing Company's biotechnology unit, who stated that while the "spires" may not be the work of any "transitory intelligence," they were nonetheless worthy of interest to scientists. Blair said, in a statement to the *Los Angeles Times*:

> If such a complex of structures were photographed on earth, the archaeologist's first order of business would be to

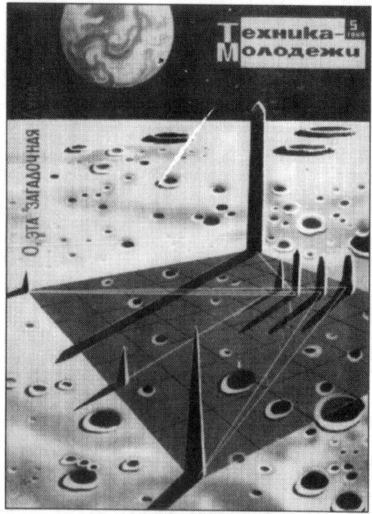

A Russian drawing of the Blair Cuspids.

inspect and excavate test trenches and thus validate whether the prospective site has archaeological significance.

Beacons for Aliens?

The great British-American zoologist, scientist and researcher Ivan T. Sanderson wrote an article about the Blair Cuspid obelisks for the August 1970 issue of *Argosy* magazine of which he was the science editor. He discussed the Blair Cuspids and also the obelisk-like objects that were photographed by the Soviet *Luna-9* probe in early 1966. The Soviet space probe *Luna-9* took some startling photographs on February 4, 1966 after the probe had landed on the Ocean of Storms, one of those dark, circular "seas" of lava on the Earth side of the Moon. The photos revealed strange towering structures that appear to be arranged in a line rather than as objects scattered randomly across the lunar surface.

Said Sanderson in 1970:

> Four years ago, Russia's *Luna-9* and America's *Orbiter-2* both photographed groups of solid structures at two widely separated locations on the lunar surface. These two groups of objects are arranged in definite geometric patterns and appear to have been placed there by intelligent

The Russian *Luna-9* took this photo of rocks spires in a direct line.

beings. Since American space officials have chosen not to publicize these findings, our readers are probably not aware of their existence.

The *Luna-9* photographs, taken on February 4, 1966, after the craft had landed in the Ocean of Storms, reveal two straight lines of equidistant stones that look like the markers along an airport runway. These circular stones are all identical, and are positioned at an angle that produces a strong reflection from the sun, which would render them visible to descending aircraft.

Upon examining the photographs, Russian scientist Dr. S. Ivanov, winner of a Laureate of the State Prize (equivalent to a Nobel Laureate) and inventor of stereo movies in the U.S.S.R., noted that a chance displacement of *Luna-9* on its horizontal axis had caused the second and third shots of the stones to be taken at slightly different angles. This double set of photographs allowed him to produce a three-dimensional stereoscopic view of the lunar "runway."

Why the *Luna-9* station changed its position between its second and third transmissions is not known. The official Russian explanation was: "Deformation of the lunar surface. The ground may have settled at the spot where the station landed, or perhaps a small stone caused the initial instability."

Whatever the reason, it was good luck for the Russian observers. "With the stereoscopic effect," reported Dr. S. Ivanov and Engineer Dr. A. Bruenko, "we can affirm that the distance between stones, one, three, two and four is equal. The stones are identical in measurement. There does not seem to be any height or elevation nearby from which the stones could have been rolled and scattered into this geometric form. The objects as seen in three-D seem to be arranged according to definite geometric laws."

The second set of photographs were taken by America's *Orbiter-2* on November 20, 1966, twenty-nine miles above the lunar surface, over the Sea of Tranquility. The photographs, of an area some 2,000 miles from the "runway" reported by the Russians on the Ocean of Storms, show what appear to be the shadows of eight pointed spires shaped like Cleopatra's Needle (the ancient Egyptian obelisk now in Central Park in New York) and the Washington Monument.

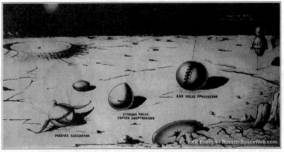

The Russian *Luna-9* landing sequence.

Since *Orbiter-2*'s cameras were pointed straight down at these spires, only their shadows are visible; but NASA stated that the sun was eleven degrees above the horizon, and from this, American space scientists estimated that the "largest protuberance" is approximately fifty feet wide at the base and from forty to seventy-five feet high.

However, the Russian scientists who examined the *Orbiter-2* photos disagreed with these American estimates, and said that the smallest of these eight apparent obelisks was "similar to an extremely large fir tree," while the largest was, by their estimation, three times the height claimed by the Americans—or as tall as a fifteen-story building!

In addition, Soviet Space Engineer Alexander Abramov has come up with a rather startling geometrical analysis of the arrangement of these objects. By calculating the angles at which they appear to be set, he asserts that they constitute an "Egyptian triangle" on the moon—a precise geometric configuration known in ancient Egypt as an *abaka*. "The distribution of these lunar objects," states Abravov, "is similar to the plan of the Egyptian pyramids constructed by Pharaohs Cheops, Chephren and Menkaura at Gizeh, near Cairo. The centers of the spires in this lunar *abaka* are arranged in precisely the same way as the apices of the three great pyramids."

What is America's position regarding the investigation of the mysterious moon objects? A high NASA authority, when questioned on what has been done by us in the four years since these objects were photographed, replied, "Yes, we know of these photographs and they were very clear, but there has been no speculation on them, and they have been filed for now."

We fail to appreciate such an attitude toward something of the first interest to our whole moon-probing endeavor; especially one which has been so openly investigated by our rivals in this effort.

However, in 1968, NASA released a very remarkable publication, the "Chronological Catalogue of Reported Lunar Events." This list includes lights, both stationary and moving, appearing on the moon and then suddenly disappearing; some perfectly circular craters that look more like domes and which are, in some cases, arranged in perfect alignment, and such phenomena as glowing mists and sudden patches of gem-colored outpourings.

John O'Neil, former science editor of the late New York *Herald Tribune* said that he observed a gigantic bridge-like structure in the Sea of Crises (on the moon) under which the sun shone when at a low angle. This was later confirmed by other astronomers. Recent literature has been crammed with descriptions of wall-like structures that form perfect squares or rhombs, of rills that look like water-eroded river beds, a grid of lines that look like roads, and even a grid of streets on the lunar surface.

Of course, many such apparently man-made structures, seen on earth from on high, later prove to be simply natural formations. At the southern edge of the Sahara Desert in Northern Nigeria is one such natural formation, which when seen from above, appears to be an enormous layout of structured walls.

And from an orbiting space craft, the Barringer Crater in Arizona looks like an artificial construction.

Despite both manned and unmanned landings on the moon, we still know very little of just what is on its surface,

The Blair Cuspids as mentioned in a 1970s comic book.

simply because of its size, the wide variety of its topography, and the still comparatively limited range of our probes.

Suppose for a moment that the earth was probed from the moon by two manned landings—let us say in Utah and East Africa. How much would we learn of the surface of the rest of the planet? Our oceans, lakes, mountain ranges and rivers might well be spotted and photographed from on high on approach, but just how many smaller things, like pyramids and obelisks, built by thinking creatures with an intelligent design in mind, might go unrecognized; or, still worse, might even be seen and filed away simply because they did not accord with what was accepted as possible by the experts at the home base.

Though very little attention has been given to the mysterious moon objects in this country, both the *Orbiter-2* obelisk photographs and the *Luna-9* runway pictures were widely published in the Soviet Union, for Russian scientists have always been extremely interested in the pursuit of any evidence of extraterrestrial life. Moreover, the question the Russian scientists are now asking about these lunar objects is whether intelligent beings could have visited our moon long ago, and erected thereupon permanent monuments and landing fields.

The question should not come as any great surprise to us because the Russians have long and consistently gone after archeological and historical evidence of superior life forms having visited this planet.

The material amassed by them in support of such an idea is now somewhat overwhelming. For instance, they claim that many Biblical stories, such as the apparent destruction of Sodom and Gomorrah by an atomic bomb, are historical accounts of such visitations. They have reported finding metallic discs, like modern recording platters, in Asiatic caves, and they have published reproductions of early Christian wall paintings from old monasteries in Yugoslavia that appear to show angels in space ships.

There are quite a number of strange material things that strongly suggest some extraterrestrial origin or influence. The Egyptian pyramids continue to puzzle scholars, who are now studying their sealed rooms with the most advanced and sensitive electromagnetic devices. One scientist, Dr.

Amr Gohed, stated officially to the London *Times* that "Either the geometry of the pyramids is in substantial error, which would affect our readings, or there is a mystery which is beyond explanation …there is some force that defies the laws of science at work in the pyramids."

The possibility of extra-terrestrial influence on the moon was put forth at a meeting of the American Rocketry Society by Dr. Carl Sagan, who said, rather simply, that: "Intelligent beings from elsewhere in the universe may have—or have had bases on the averted side of our moon."

Why, we may well ask, did man ever start making obelisks anyway? It's a very tough job and seemingly purposeless. Is the origin of the obelisks on this earth, and those on the moon, the same? Could both be ancient markers originally erected by alien space travelers for guidance of later arrivals?

Sanderson asks the important questions that I am struggling to answer in this book: why were obelisks made by ancient civilizations (we don't know) and are the obelisks on the Moon the same as those on Earth?

Sanderson theorizes that they might have been erected by ancient astronauts—extraterrestrials—as beacons for guidance of other space travelers. This seems to be on the right track. And beacons for space travelers would need to have some energy associated with them and granite stone that is infused with tiny quartz crystals can provide an energy signature. It can also be an antenna or power transmitter.

We haven't learned a whole lot more about the alignments of stones and obelisks on the Moon. It would seem to be a good place to land another probe, or possibly put a space base. This one comes with a pre-set formation of monoliths, just like in *2001: A Space Odyssey*. Have we finally come to face the monoliths on the Moon as the space message that somebody may have been intended them to be?

We haven't got the final analysis of these formations; we can however look at a few other unusual objects on the Moon and even on Mars.

Structures on the Moon
Starting in the 1970s a number of researchers were intrigued by Sanderson's article in *Argosy* and books began to appear such as

310

George Leonard's 1977 book *Somebody Else is On the Moon.*[58] In that book Leonard reveals that the Soviet magazine *Technology of Youth* gave an extensive report on the Blair Cuspid's, calling them "stone markers" which were unquestionably "planned structures," and suggested that these "pointed pyramids" were not natural formations but definitely artificial structures of alien origin.

As we have seen, after examining the photographs of the Blair Cuspids, Dr. S. Ivanov calculated from the shadows cast by the obelisk-like structures that at least one of them was about fifteen stories high. Says Leonard about Ivanov and *Luna-9* photos that show a different configuration of stones:

> Ivanov, who is also the inventor of stereo movies in the Soviet Union, pointed out that by luck—perhaps the space probe landed on a spot where the ground had settled, or set down upon a small stone or rough spot—"a chance displacement of *Luna-9* on its horizontal axis had caused the stones to be taken at slightly different angles." This double set of photographs allowed him to produce a three-dimensional stereoscopic view of the lunar "runway."
>
> The result of this bit of good fortune, as Ivanov reports, was that the stereoscopic effect enabled scientists to figure the distances between the spires. They found, much to their surprise, that they were spaced at regular intervals. Moreover, calculations confirmed that the spires themselves were identical in measurement. This discovery must be heralded as among the most important discoveries made by either the American or Soviet space program. But, strangely enough, for the most part they have been ignored. As we shall soon see, other discoveries, equally as important, have been covered up by our own space agency. In fact, Art Rosenblum, head of the Aquarian Research Foundation, who says he learned of the Soviet discovery from Lynn Schroeder and Sheila Ostrander, the authors of *Psychic Discoveries Behind the Iron Curtain,* before their work was published in America, claims they indicated that authorities at NASA "were not at all happy about its publication." Why not? What is NASA trying to hide? asks Rosenblum. (Arthur Rosenblum, *Unpopular Science*, Running Press, 1974.)[58]

Leonard also speculates that the obelisks such as the Blair

Cuspids were markers for underground bases. Says Leonard:

> Dr. Sanderson speculated: "Is the origin of the obelisks on the Earth and those on the Moon the same? Could both be ancient markers originally erected by alien space travelers for guidance of late arrivals?" He pointed out that it seems hard to understand why man ever started making obelisks anyway, since it is a very difficult job and seemingly purposeless. Or did obelisks have a purpose other than Earthly? Could these spire-like structures actually be signal spots for the coming and going of spaceships, as some speculated? Not marking the landing on outer Moon bases but for underground, hidden bases located inside the Moon?
>
> Intriguingly, on the edge of this same Sea of Storms is a strange opening that leads down into the Moon. Dr. H.P. Wilkins, one of the world's leading lunar experts before his untimely death a few years ago, was convinced that extensive hollow areas did exist inside the Moon, perhaps in the form of caverns, and that these were connected to the surface by huge holes or pits. He discovered such an opening himself—a huge round hole inside the crater Cassini A. This crater is one and a half miles across, and the opening leading down into the Moon is over 600 feet across—more than two football fields laid end to end. Wilkins writes in his definitive work, *Our Moon*: "Its inside is as smooth as glass with a deep pit or plughole, about 200 yards across at the center." [58]

Leonard says that hundreds, or even thousands, of UFOs have been seen on or around the surface of the Moon, and that a concentration of them has been spotted in the area known as the Sea of Storms with its large opening. Could the UFOs be coming and going through this huge aperture or one like it?

Leonard comments on the strange pyramidal, obelisk-like structures in the Sea of Tranquility. He says that the cuspids have probably been closely examined by NASA, since NASA astronauts went to the very same Sea of Tranquility on their first trip to the Moon. Leonard says that "the results have never been released and probably never will be unless enough public pressure can be applied on our government." [58]

Leonard also mentions an anomaly that is inside of "Ranger 7 crater." The controversial "Ranger 7 crater" was a crater seen in the last photo that *Ranger 7* took of the Moon as it was about to crash into the surface. That photo showed a cluster of objects inside a crater and another crater with a pyramid in the center:

> The first phase of our Moon exploration, the Ranger series, returned thousands of photos of the Moon's surface— close-up pictures that revealed a great deal about the surface of this strange satellite. One of these taken by *Ranger 7* just before it smashed into the Moon has produced a storm of controversy. The photo was taken about three miles on the last leg of its crash dive and appears to show some objects inside a crater. NASA officials claim they are just "a cluster of rocks." Other investigators are not so sure.
>
> No one knows for sure but some investigators like Riley Crabb of the Borderland Science Research Organization claims that their circular symmetry indicates that whatever they are, certainly they are "intelligently constructed."
>
> Crabb believes that this conclusion is "confirmed by the sharp, straight black shadow cast" by one of two "brilliant white shafts" inside the crater. He points out that this August 10, 1964, photo published in *Missiles and Rockets* magazine (p. 22), shows "clearly outlined in between the bases of the two shafts a perfect circle, perhaps 40 or 50 feet across. The hole itself is pitch black, as though it led into the interior

The Russian *Zond 3* took this photo of a gigantic tower on the Moon. **313**

of the moon; but the edges are bright, like the edges of a gigantic bubble or lens."

Leonard and others, including myself, have written or edited books on anomalous "extraterrestrial" structures. Mike Bara in his book *Ancient Aliens on the Moon*, discusses the little-known Russian *Zond 3* picture of a 20-mile high tower located at a spot on the Moon just west of the area known as the Oceanus Procellarum. Says Bara:

> It was thought in the West that *Zond 3* was intended as a companion mission to *Zond 2*, and that perhaps the two spacecraft were intended to meet up in orbit around Mars. However, for reasons unknown, *Zond 3* missed its launch window in 1964 and was instead launched in 1965 on a Mars trajectory, even though Mars was no longer in the same location. As a result, it was useful only as a guidance/telemetry test vehicle, and its only real data would be acquired in the flyby of the Moon that was needed to place it on a Mars orbit trajectory.
>
> As it passed by the Moon with a closest approach distance of some 5,716 miles, the *Zond 3* spacecraft snapped 23 photos and took 3 spectral images of the lunar far side. Centered mostly over the Mare Orientale impact basin, most of the images were fairly non-descript. But two stood out as completely remarkable. The first image (frame 25) was first published in defense contractor TRW's *Solar System Log* magazine in 1967. It shows yet *another* tower, sticking straight up from the lunar surface along the visible limb of the Moon.
>
> This remarkable object is actually anchored somewhere over the horizon, near the towering structures on the western edge of Oceanus Procellarum that were visible from the ground by *Apollos 12* and *14*. Again, by definition such an object (which is at least 20 miles high) *has* to be artificial because no natural object could be standing upright against the incessant meteoric rain of the last 4.5 billion years of the Moon's existence.
>
> Close-up enhancements of the image shows that not only does the *Zond 3* tower appear to be anchored at a point over the horizon, there are some odd and very geometric

looking objects next to it. Again, a natural looking lunar horizon should be very smooth, not broken up as the area around the tower is, and of course nothing like the tower should be there at all...

The next *Zond 3* image, taken thirty-four seconds later, had no tower visible at all, indicating it had, by that time, slipped over the lunar horizon due to the fast moving Soviet spacecraft's motion and direction. After the tower vanished out of view, it was immediately replaced by an equally anomalous feature, found on image frame 28.

The second amazing *Zond 3* shot was originally found in an official NASA publication, *Exploring Space with a Camera* (NASA SP-168, 1968), but is now available in high-resolution on the internet.

Located on the lunar horizon approximately a thousand miles further to the south, a large, eroded dome-like structure was plainly visible in the lower right corner of the image.

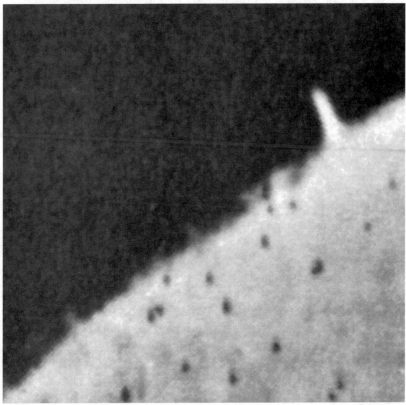

A close-up of the gigantic tower on the Moon photographed by the *Zond 3*.

Again, this "Zond 3 dome" extended several miles above the airless lunar horizon against black space. And, like the earlier "Zond tower," neatly aligned with the local vertical.

A close-up of this second *Zond 3* dome reveals a significant amount of deterioration, undoubtedly due to long term exposure to the effects of meteor erosion. That said, the outline of a very geometric, structural building extending miles above the Moon is still clearly defined. Above the dome, the smaller remnants of what appears to be a latticework type structure are visible under enhancement. Based on this, I suspect that what *Zond 3* captured was in fact the battered remains of a watch crystal type of dome just beneath the larger protecting scaffolding we've seen in other places on the lunar surface. Such an arrangement would provide an ideal engineering solution to the long-term issues involved with lunar habitation. Without an atmosphere to protect it from even the most mundane, everyday kinds of meteor strikes, a lunar base would stand little chance of surviving beyond a few years. But with a multilayered, miles-high scaffolding structure at the top and then smaller reinforcing "watch crystal" domes underneath, a lunar base might conceivably survive for hundreds of thousands if not millions of years.

But there is ground truth, and then there is ground truth. The next obvious question is if these towering structures were all around the astronauts as they descended to and then explored the lunar surface, why didn't they *see* them? And if they did, why then did they not comment on them?

The first question is actually the easier one to answer. But the second one is the more revealing. After all, what if they did see these structures, but somehow just *forgot* that they did?

In his last sentence Bara is suggesting here, as others have suggested, that the Apollo astronauts witnessed a number of unusual structures and other objects while they were on the Moon or flying around it. But when they returned back to the USA they were debriefed and hypnotized by trained professionals. The astronauts were told to forget that they had seen towers, obelisks, pyramids and domes and the astronauts forgot about them. They did not talk about such things as they genuinely could not remember them.

316

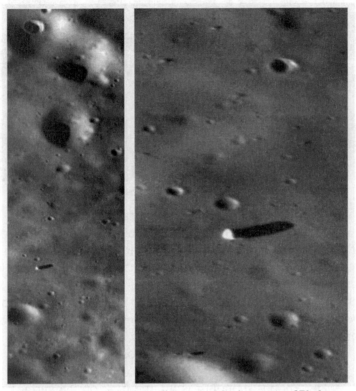

Different photos of the monolith on the Martian moon of Phobos.

Bara has also said on a number of occasions that he personally spoke at a NASA convention with one of the hypnotists who said she had hypnotized some of the astronauts when they returned from the Moon. When Bara asked her what had happened during those hypnotic sessions and what did the astronauts say?—she replied that she did not know because she had been hypnotized herself after she had performed hypnosis on the astronauts. She knew that she had performed hypnotic sessions on the astronauts but could not remember those sessions herself! How intricately layered our onion of a scenario is as we explore the topic of obelisks on the Moon.

What we need to ask at this point is who made these structures on the Moon? How old are they? Do they mark an underground base for extraterrestrials that are still occupying these facilities? Are they antennas and beacons for extraterrestrials? Are obelisks on the earth serving the same function or did they do so in the past? The answers here seem to be a resounding yes!

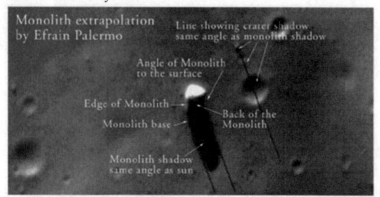

An analysis of the Phobos monolith by Efrain Palermo.

A Final Look at the Monoliths on Mars

We must finally look at the astounding photos taken of the Martian moon Phobos which seem to show towering monoliths on its surface. The Phobos monolith is an object in photographs taken by the *Mars Global Surveyor* (such as MOC Image 55103) in 1998. Studies of photos by researcher Efrain Palermo show that it appears to be a massive obelisk with vertical sides and a pyramidion on the summit.

Wikipedia has an entry for the Phobos monolith that says:

> The Phobos monolith is a large rock on the surface of Mars's moon Phobos. It is a boulder about 85 m (279 ft) across and 90 m (300 ft) tall. A monolith is a geological feature consisting of a single massive piece of rock. Monoliths also occur naturally on Earth, but it has been suggested that the Phobos monolith may be a piece of impact ejecta. The monolith is a bright object near Stickney crater, described as a "building sized" boulder, which casts a prominent shadow. It was discovered by Efrain Palermo, who did extensive surveys of Martian probe imagery, and later confirmed by Lan Fleming, an imaging sub-contractor at NASA Johnson Space Center. Lan Fleming considered the possibility that the Phobos monolith may be artificial and not a geological feature or rock.
>
> The general vicinity of the monolith is a proposed landing site by Optech and the Mars Institute, for an unmanned mission to Phobos known as PRIME (Phobos Reconnaissance and International Mars Exploration). The PRIME mission would be composed of an orbiter and lander,

and each would carry four instruments designed to study various aspects of Phobos' geology. At present, PRIME has not been funded and does not have a projected launch date. Former astronaut Buzz Aldrin has spoken about the Phobos monolith and his support for a mission to Phobos.

Palermo also noticed a group of smaller obelisk-type objects on a crater to the west of the Phobos monolith. These objects also cast long shadows and appear to be tall, slender objects with straight sides. Are they a cluster of obelisks? Does the Martian moon of Phobos have a nearly 100 meter-high monolith with a pointed top like an obelisk or pyramidion? It would seem so.

Another monolith was discovered by NASA's *Mars Reconnaissance Orbiter* (MRO). The orbiter was launched August 12, 2005, and attained Martian orbit on March 10, 2006. In November 2006, after five months of aerobraking, it entered its final science orbit and began its primary science phase. As MRO entered orbit, it joined five other active spacecraft that were either in orbit or on the planet's surface: *Mars Global Surveyor, Mars Express, 2001 Mars Odyssey*, and the two Mars Exploration Rovers (*Spirit* and

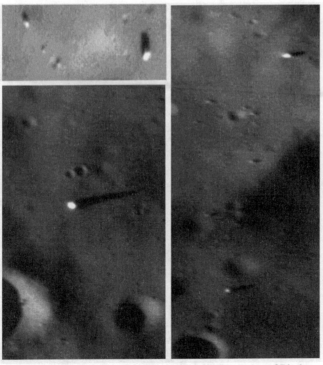

Various photos of the monolith on the Martian moon of Phobos. **319**

Opportunity); at the time, this set a record for the most operational spacecraft in the immediate vicinity of Mars. During this time it photographed a rectangular monolith on the surface of the planet. Says the brief Wikipedia entry:

> The Mars monolith is a rectangular object (possibly a boulder) discovered on the surface of Mars. It is located near the bottom of a cliff, from which it likely fell. The Mars Reconnaissance Orbiter took pictures of it from orbit, roughly 180 miles (300 km) away. It is estimated to measure about 5 meters wide. Around the same time the Phobos monolith made international news.

A picture tells a thousand words and so we should look at the photo and see if this looks like an artificial monolith placed on the surface of Mars. Indeed, it does look like that and has a distinct likeness to the monolith in Arthur C. Clarke's *2001: A Space Odyssey*. Perhaps he was a visionary with some inside knowledge.

As we explore our solar system we may find more monolith and obelisk-type structures. Many of them are bound to be artificial. Who built them and what was their purpose?

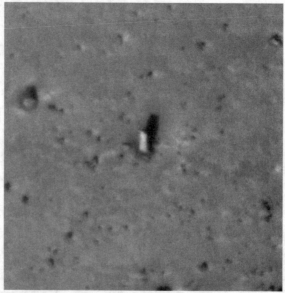

A photograph of the Mars monolith.

BIBLIOGRAPHY & FOOTNOTES

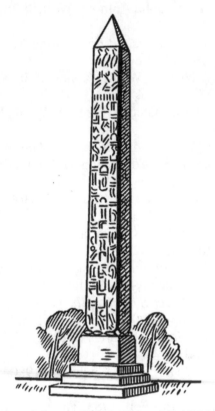

1. *The Obelisks of Egypt*, Labib Habachi, 1984, American University Press, Cairo.
2. *The Magic of Obelisks*, Peter Tompkins, 1981, Harper and Row Publishers. New York.
3. *Ancient Mysteries, Modern Visions*, Philip Callahan, 1984, Acres USA, Kansas City, MO.
4. *The Sign and the Seal*, Graham Hancock, 1992, Crown Publishers, New York.
5. *Secrets of the Great Pyramid*, Peter Tompkins, 1971, Harper & Row, New York.
6. *The Great Pyramid*, Piazzi Smyth, 1880, Bell Publishing Co. New York.
7. *Obelisk: A History*, Curran, Grafton, Long and Weiss, 2000, Burndy Library, Cambridge, MA.

8. *The Problem of the Obelisks*, Reginald Engelbach, 1923, George H. Doran, New York.
9. *Ancient Egyptian Metallurgy*, H. Garland and C.O. Bannister, 1927, London.
10. *Serpent in the Sky*, John Anthony West, 1978, Alfred Knopf, New York.
11. *A Traveller's Key to Ancient Egypt*, John Anthony West, 1984, Alfred Knopf, New York.
12. *The Bible As History*, Werner Keller, 1956, William Morrow, NYC.
13. *Kebra Nagast*, translated by Sir E. A. Wallis Budge, 1932, Dover, London.
14. *In Search of Quetzalcoatl (In Search of the White God)*, Pierre Honoré, 1961, Adventures Unlimited Press, Kempton, IL (Reprint).
15. *Egypt Before the Pharaohs*, Michael A. Hoffman, 1979, Alfred Knopf, New York.
16. *The Giza Power Plant*, Christopher Dunn, 1998, Bear & Company, Rochester, VT.
17. *Lost Technologies of Ancient Egypt*, Christopher Dunn, 2010, Bear & Company, Rochester, VT.
18. *Atlantis & the Power System of the Gods*, David Childress and Bill Clendenon, 2002, AUP, Kempton, IL.
19. *The Bermuda Triangle*, Charles Berlitz, 1974, Doubleday, Garden City, NJ.
20. *Gateway to Oblivion*, Hugh Cochrane, 1980, Anchor Press, London.
21. *Saga America*, Barry Fell, 1983, Times Books, NYC.
22. *Alexandria: A History & a Guide*, E.M. Forster, 1922, Morris, Alexandria.
23. *Technology In the Ancient World*, Henry Hodges, 1970, Marboro Books, London.
24. *We Are Not the First*, Andrew Tomas, 1971, Souvenir Press, London.
25. *The Phoenicians*, Gerhard Herm, 1975, William Morrow & Co. New York.
26. *Ancient Ethiopia*, David W. Phillipson, 1998, British Museum Press, London.
27. *The Bible As History*, Werner Keller, 1956, William Morrow & Co. New York.
28. *Stonehenge Decoded*, Gerald Hawkins, 1965, Doubleday, NY.

29. *Anti-Gravity & the World Grid*, David Hatcher Childress, 1987, AUP, Kempton, IL.
30. *Technology of the Gods*, David Hatcher Childress, 1999, AUP, Kempton, IL.
31. *Ancient Technology in Peru and Bolivia*, 2014, AUP, Kempton, IL.
32. *Remarkable Luminous Phenomena in Nature*, William Corliss, 2001, Sourcebook Project, Glen Arm, MD.
33. *Ancient Man: A Handbook of Puzzling Artifacts*, William Corliss, 1978, Sourcebook Project, Glen Arm, MD.
34. *Ball Lightning Explanation Leading to Clean Energy*, Clint Seward, 2011, Amazon.com
35. *Atlantis & the Power System of the Gods*, Bill Clendenon and David Hatcher Childress, 2002, AUP, Kempton, IL.
36. *Round Towers of Ireland,* Henry O'Brien, 1834, reprinted as *Round Towers of Atlantis,* 2002, by Adventures Unlimited Press.
37. *Vimana: Flying Machines of the Ancients*, David Hatcher Childress, 2013, AUP, Kempton, IL.
38. *Secrets of the Lost Races*, Rene Noorbergen, 1980, Bobbs-Merrill Co. New York
39. *Baalbek*, Friedrich Ragette, 1980, Chatto & Windus, London.
40. *The Incas*, Garcilaso de la Vega, 1961, Penguin Books (first published in 1608).
41. *A Dweller on Two Planets,* Phylos the Thibetan, 1899, Reprinted AUP, Kempton, IL.
42. *An Earth Dweller Returns*, Phylos the Thibetan, 1940, Borden Press, Los Angeles.
43. *What the Bible Really Says*, Manfred Barthel, 1982, Souvenir Press, London.
44. *Mysteries of Forgotten Worlds*, Charles Berlitz, 1972, Doubleday, New York.
45. *6000 Years of Seafaring*, Orville Hope, 1983, Hope Associates, Gastonia, NC.
46. *Ships and Seamanship in the Ancient World*, Lionel Casson, 1971, Princeton University Press, Princeton, NJ.
47. *Easy Journey to Other Planets,* Swami Prabhupada, 1970, Bhaktivedanta Book Trust (ISKCON), Los Angeles.
48. *Egyptian Myth & Legend*, Donald Mackenzie, 1907, Bell Publishers, New York.

49. *The Encyclopedia of Moon Mysteries*, Constance Victoria Briggs, 2019, AUP, Kempton, IL.

50. *War in Ancient India*, Ramachandra Dikshitar, 1944, Motilal Banarsidass, Delhi.

51. *Inca Architecture and Construction at Ollantaytambo*, Jean-Pierre Protzen, 1993, Oxford University Press.

52. *The Giza Death Star*, Joseph P. Farrell, 2001, AUP, Kempton, IL.

53. *The Yahweh Encounters*, Ann Madden Jones 1995, Sandbird Publishing, Chapel Hill, NC.

54. *Laserbeams from Star Cities*, Robyn Collins, 1971, Sphere Books, London.

55. *Secret Cities of Old South America*, Harold Wilkins, 1952, London, reprinted 1998, Adventures Unlimited Press, Kempton, Illinois.

56. *Mercury: UFO Messenger of the Gods,* William Clendenon, 1990, Adventures Unlimited Press, Kempton, IL.

57. *Extraterrestrial Archeology*, David Childress, ed. 1994, AUP, Kempton, Illinois.

58. *Somebody Else is On the Moon*, Geoge Leonard, 1977, Prentice-Hall, New Jersey.

59. *Ancient Aliens on the Moon*, Mike Bara, 2012, AUP, Kempton, IL.

60. *Ancient Aliens on Mars*, Mike Bara, 2013, AUP, Kempton, IL.

61. *Investigating the Unexplained*, Ivan T. Sanderson, 1972, Prentice Hall, Englewood Cliffs, NJ.

62. *Atlantis: From Legend to Discovery*, Andrew Tomas, 1972, Robert Laffont, Paris. (Sphere Books, 1973, London).

63. *The Shadow of Atlantis*, Colonel A. Braghine, 1940, London reprinted 1996, Adventures Unlimited Press, Kempton, IL.

63. *Ancient Egyptian Metallurgy,* H. Garland and C.O. Bannister, 1927, London.

64. *Anti-Gravity & the World Grid,* D.H. Childress, 1987, Adventures Unlimited Press, Stelle, Illinois.

65. *Edgar Cayce On Atlantis,* Hugh Lynn Cayce, 1968, A.R.E., Virginia Beach, VA.

66. *Anti-Gravity & the Unified Field*, D.H. Childress, ed. 1990, AUP, Kempton, Illinois.
67. *The Tesla Papers,* D.H. Childress, ed. 1999, AUP, Kempton, Illinois.
68. *The Fantastic Inventions of Nikola Tesla,* D.H. Childress, ed. 1993, AUP, Kempton, Illinois.
69. *Quest For Zero Point Energy,* Moray B. King, 2002, AUP, Kempton, Illinois.
70. *Anti-Gravity & the Unified Field*, Edited by David Hatcher Childress, 1990, Adventures Unlimited Press, Kempton, Illinois.
71. *Pyramid Rising*, Zach Royer, 2016, Kahuna Research, Kona, HI.
72. *Picture-Writing of the American Indians*, Garrick Mallory, 1889, Smithsonian Institute, Reprinted by Dover Books.
73. *The Secret Life of Plants,* Christopher Bird and Peter Thompkins, 1984, Harper & Row, New York.
74. *Phoenicians*, Glen Markoe, 2000, University of California Press, Berkeley.
75. *Magic and Mystery In Tibet*, Alexandra David-Neel, 1935, AUP, Kempton, IL.
76. *The Sphinx and the Megaliths*, John Ivimy, 1974, Sphere Books, London.
77. *The Pyramids: An Enigma Solved*, Joseph Davidovits & Margie Morris, 1988, Hippocrene Books, New York.
78. *Irish Druids and Old Irish Religions,* James Bonwick, 1894, reprinted by Dorset Press, 1986, New York.
79. *Ancient Monuments and Stone Circles*, Les Morres, 2000, Frith Books, Salisbury, UK.
80. *The Cosmic Serpent*, Meremy Narby, 1998, Jeremy Tarcher, New York.
81. *Lost Cities & Ancient Mysteries of South America*, David Hatcher Childress, 1987, AUP, Kempton, Illinois.

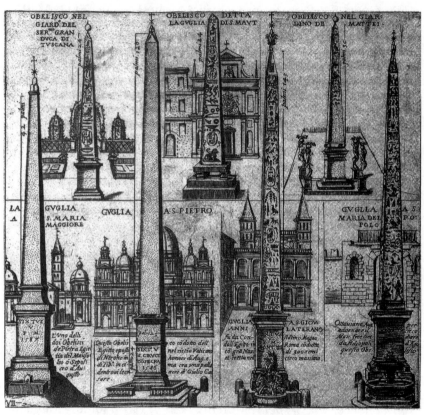

An old print of the various obelisks erected in Rome.

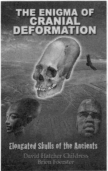
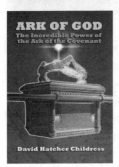

YETIS, SASQUATCH & HAIRY GIANTS
By David Hatcher Childress
Childress takes the reader on a fantastic journey across the Himalayas to Europe and North America in his quest for Yeti, Sasquatch and Hairy Giants. Childress begins with a discussion of giants and then tells of his own decades-long quest for the Yeti in Nepal, Sikkim, Bhutan and other areas of the Himalayas, and then proceeds to his research into Bigfoot, Sasquatch and Skunk Apes in North America. Chapters include: The Giants of Yore; Giants Among Us; Wildmen and Hairy Giants; The Call of the Yeti; Kanchenjunga Demons; The Yeti of Tibet, Mongolia & Russia; Bigfoot & the Grassman; Sasquatch Rules the Forest; Modern Sasquatch Accounts; more. Includes a 16-page color photo insert of astonishing photos!
360 pages. 5x9 Paperback. Illustrated. Bibliography. Index. $18.95. Code: YSHG

SECRETS OF THE HOLY LANCE
The Spear of Destiny in History & Legend
by Jerry E. Smith
Secrets of the Holy Lance traces the Spear from its possession by Constantine, Rome's first Christian Caesar, to Charlemagne's claim that with it he ruled the Holy Roman Empire by Divine Right, and on through two thousand years of kings and emperors, until it came within Hitler's grasp—and beyond! Did it rest for a while in Antarctic ice? Is it now hidden in Europe, awaiting the next person to claim its awesome power? Neither debunking nor worshiping, *Secrets of the Holy Lance* seeks to pierce the veil of myth and mystery around the Spear.
312 PAGES. 6x9 PAPERBACK. ILLUSTRATED. $16.95. CODE: SOHL

THE CRYSTAL SKULLS
Astonishing Portals to Man's Past
by David Hatcher Childress and Stephen S. Mehler
Childress introduces the technology and lore of crystals, and then plunges into the turbulent times of the Mexican Revolution form the backdrop for the rollicking adventures of Ambrose Bierce, the renowned journalist who went missing in the jungles in 1913, and F.A. Mitchell-Hedges, the notorious adventurer who emerged from the jungles with the most famous of the crystal skulls. Mehler shares his extensive knowledge of and experience with crystal skulls. Having been involved in the field since the 1980s, he has personally examined many of the most influential skulls, and has worked with the leaders in crystal skull research. Color section.
294 pages. 6x9 Paperback. Illustrated. $18.95. Code: CRSK

THE LAND OF OSIRIS
An Introduction to Khemitology
by Stephen S. Mehler
Was there an advanced prehistoric civilization in ancient Egypt? Were they the people who built the great pyramids and carved the Great Sphinx? Did the pyramids serve as energy devices and not as tombs for kings? Chapters include: Egyptology and Its Paradigms; Khemitology—New Paradigms; Asgat Nefer—The Harmony of Water; Khemit and the Myth of Atlantis; The Extraterrestrial Question; more. Color section.
272 PAGES. 6x9 PAPERBACK. ILLUSTRATED . $18.95. CODE: LOOS

VIMANA:
Flying Machines of the Ancients
by David Hatcher Childress
According to early Sanskrit texts the ancients had several types of airships called vimanas. Like aircraft of today, vimanas were used to fly through the air from city to city; to conduct aerial surveys of uncharted lands; and as delivery vehicles for awesome weapons. David Hatcher Childress, popular *Lost Cities* author, takes us on an astounding investigation into tales of ancient flying machines. In his new book, packed with photos and diagrams, he consults ancient texts and modern stories and presents astonishing evidence that aircraft, similar to the ones we use today, were used thousands of years ago in India, Sumeria, China and other countries. Includes a 24-page color section.
408 Pages. 6x9 Paperback. Illustrated. $22.95. Code: VMA

THE LOST WORLD OF CHAM
The Trans-Pacific Voyages of the Champa
By David Hatcher Childress

The mysterious Cham, or Champa, peoples of Southeast Asia formed a megalith-building, seagoing empire that extended into Indonesia, Tonga, and beyond—a transoceanic power that reached Mexico and South America. The Champa maintained many ports in what is today Vietnam, Cambodia, and Indonesia and their ships plied the Indian Ocean and the Pacific, bringing Chinese, African and Indian traders to far off lands, including Olmec ports on the Pacific Coast of Central America. opics include: Cham and Khem: Egyptian Influence on Cham; The Search for Metals; The Basalt City of Nan Madol; Elephants and Buddhists in North America; The Olmecs; The Cham in Colombia; The Cham and Lake Titicaca; Easter Island and the Cham; the Magical Technology of the Cham; tons more. 24-page color section.

328 Pages. 6x9 Paperback. Illustrated. $22.00 Code: LPWC

MIND CONTROL, OSWALD & JFK
Introduction by Kenn Thomas

In 1969 the strange book *Were We Controlled?* was published which maintained that Lee Harvey Oswald was a special agent who was also a Mind Control subject who had received an implant in 1960. Thomas examines the evidence that Oswald had been an early recipient of the Mind Control implant technology and this startling role in the JFK Assassination. Also: the RHIC-EDOM Mind Control aspects concerning the RFK assassination and the history of implant technology.

256 Pages. 6x9 Paperback. Illustrated. $16.00. Code: MCOJ

INSIDE THE GEMSTONE FILE
Howard Hughes, Onassis & JFK
By Kenn Thomas & David Childress

Here is the low-down on the most famous underground document ever circulated. Photocopied and distributed for over 20 years, the Gemstone File is the story of Bruce Roberts, the inventor of the synthetic ruby widely used in laser technology today, and his relationship with the Howard Hughes Company and ultimately with Aristotle Onassis, the Mafia, and the CIA. Hughes kidnapped and held a drugged-up prisoner for 10 years; Onassis and his role in the Kennedy Assassination; how the Mafia ran corporate America in the 1960s; more.

320 Pages. 6x9 Paperback. Illustrated. $16.00. Code: IGF

ADVENTURES OF A HASHISH SMUGGLER
by Henri de Monfreid

Nobleman, writer, adventurer and inspiration for the swashbuckling gun runner in the *Adventures of Tintin*, Henri de Monfreid lived by his own account "a rich, restless, magnificent life" as one of the great travelers of his or any age. The son of a French artist who knew Paul Gaugin as a child, de Monfreid sought his fortune by becoming a collector and merchant of the fabled Persian Gulf pearls. He was then drawn into the shadowy world of arms trading, slavery, smuggling and drugs. Infamous as well as famous, his name is inextricably linked to the Red Sea and the raffish ports between Suez and Aden in the early years of the twentieth century. De Monfreid (1879 to 1974) had a long life of many adventures around the Horn of Africa where he dodged pirates as well as the authorities.

284 Pages. 6x9 Paperback. $16.95. Illustrated. Code AHS

TECHNOLOGY OF THE GODS
The Incredible Sciences of the Ancients
by David Hatcher Childress

Childress looks at the technology that was allegedly used in Atlantis and the theory that the Great Pyramid of Egypt was originally a gigantic power station. He examines tales of ancient flight and the technology that it involved; how the ancients used electricity; megalithic building techniques; the use of crystal lenses and the fire from the gods; evidence of various high tech weapons in the past, including atomic weapons; ancient metallurgy and heavy machinery; the role of modern inventors such as Nikola Tesla in bringing ancient technology back into modern use; impossible artifacts; and more.

356 pages. 6x9 Paperback. Illustrated. $16.95. code: TGOD

THE ANTI-GRAVITY HANDBOOK
edited by David Hatcher Childress

The new expanded compilation of material on Anti-Gravity, Free Energy, Flying Saucer Propulsion, UFOs, Suppressed Technology, NASA Cover-ups and more. Highly illustrated with patents, technical illustrations and photos. This revised and expanded edition has more material, including photos of Area 51, Nevada, the government's secret testing facility. This classic on weird science is back in a new format!

230 PAGES. 7x10 PAPERBACK. ILLUSTRATED. $16.95. CODE: AGH

ANTI–GRAVITY & THE WORLD GRID

Is the earth surrounded by an intricate electromagnetic grid network offering free energy? This compilation of material on ley lines and world power points contains chapters on the geography, mathematics, and light harmonics of the earth grid. Learn the purpose of ley lines and ancient megalithic structures located on the grid. Discover how the grid made the Philadelphia Experiment possible. Explore the Coral Castle and many other mysteries, including acoustic levitation, Tesla Shields and scalar wave weaponry. Browse through the section on anti-gravity patents, and research resources.

274 PAGES. 7x10 PAPERBACK. ILLUSTRATED. $14.95. CODE: AGW

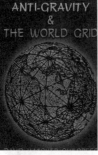

ANTI–GRAVITY & THE UNIFIED FIELD
edited by David Hatcher Childress

Is Einstein's Unified Field Theory the answer to all of our energy problems? Explored in this compilation of material is how gravity, electricity and magnetism manifest from a unified field around us. Why artificial gravity is possible; secrets of UFO propulsion; free energy; Nikola Tesla and anti-gravity airships of the 20s and 30s; flying saucers as superconducting whirls of plasma; anti-mass generators; vortex propulsion; suppressed technology; government cover-ups; gravitational pulse drive; spacecraft & more.

240 PAGES. 7x10 PAPERBACK. ILLUSTRATED. $14.95. CODE: AGU

THE TIME TRAVEL HANDBOOK
A Manual of Practical Teleportation & Time Travel
edited by David Hatcher Childress

The Time Travel Handbook takes the reader beyond the government experiments and deep into the uncharted territory of early time travellers such as Nikola Tesla and Guglielmo Marconi and their alleged time travel experiments, as well as the Wilson Brothers of EMI and their connection to the Philadelphia Experiment—the U.S. Navy's forays into invisibility, time travel, and teleportation. Childress looks into the claims of time travelling individuals, and investigates the unusual claim that the pyramids on Mars were built in the future and sent back in time. A highly visual, large format book, with patents, photos and schematics. Be the first on your block to build your own time travel device!

316 PAGES. 7x10 PAPERBACK. ILLUSTRATED. $16.95. CODE: TTH

ANCIENT ALIENS ON THE MOON
By Mike Bara
What did NASA find in their explorations of the solar system that they may have kept from the general public? How ancient really are these ruins on the Moon? Using official NASA and Russian photos of the Moon, Bara looks at vast cityscapes and domes in the Sinus Medii region as well as glass domes in the Crisium region. Bara also takes a detailed look at the mission of Apollo 17 and the case that this was a salvage mission, primarily concerned with investigating an opening into a massive hexagonal ruin near the landing site. Chapters include: The History of Lunar Anomalies; The Early 20th Century; Sinus Medii; To the Moon Alice!; Mare Crisium; Yes, Virginia, We Really Went to the Moon; Apollo 17; more. Tons of photos of the Moon examined for possible structures and other anomalies.
248 Pages. 6x9 Paperback. Illustrated.. $19.95. Code: AAOM

ANCIENT ALIENS ON MARS
By Mike Bara
Bara brings us this lavishly illustrated volume on alien structures on Mars. Was there once a vast, technologically advanced civilization on Mars, and did it leave evidence of its existence behind for humans to find eons later? Did these advanced extraterrestrial visitors vanish in a solar system wide cataclysm of their own making, only to make their way to Earth and start anew? Was Mars once as lush and green as the Earth, and teeming with life? Chapters include: War of the Worlds; The Mars Tidal Model; The Death of Mars; Cydonia and the Face on Mars; The Monuments of Mars; The Search for Life on Mars; The True Colors of Mars and The Pathfinder Sphinx; more. Color section.
252 Pages. 6x9 Paperback. Illustrated. $19.95. Code: AMAR

ANCIENT ALIENS ON MARS II
By Mike Bara
Using data acquired from sophisticated new scientific instruments like the Mars Odyssey THEMIS infrared imager, Bara shows that the region of Cydonia overlays a vast underground city full of enormous structures and devices that may still be operating. He peels back the layers of mystery to show images of tunnel systems, temples and ruins, and exposes the sophisticated NASA conspiracy designed to hide them. Bara also tackles the enigma of Mars' hollowed out moon Phobos, and exposes evidence that it is artificial. Long-held myths about Mars, including claims that it is protected by a sophisticated UFO defense system, are examined. Data from the Mars rovers Spirit, Opportunity and Curiosity are examined; everything from fossilized plants to mechanical debris is exposed in images taken directly from NASA's own archives.
294 Pages. 6x9 Paperback. Illustrated. $19.95. Code: AAM2

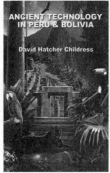

ANCIENT TECHNOLOGY IN PERU & BOLIVIA
By David Hatcher Childress
Childress speculates on the existence of a sunken city in Lake Titicaca and reveals new evidence that the Sumerians may have arrived in South America 4,000 years ago. He demonstrates that the use of "keystone cuts" with metal clamps poured into them to secure megalithic construction was an advanced technology used all over the world, from the Andes to Egypt, Greece and Southeast Asia. He maintains that only power tools could have made the intricate articulation and drill holes found in extremely hard granite and basalt blocks in Bolivia and Peru, and that the megalith builders had to have had advanced methods for moving and stacking gigantic blocks of stone, some weighing over 100 tons.
340 Pages. 6x9 Paperback. Illustrated.. $19.95 Code: ATP

HESS AND THE PENGUINS
The Holocaust, Antarctica and the Strange Case of Rudolf Hess
By Joseph P. Farrell

Farrell looks at Hess' mission to make peace with Britain and get rid of Hitler—even a plot to fly Hitler to Britain for capture! How much did Göring and Hitler know of Rudolf Hess' subversive plot, and what happened to Hess? Why was a doppleganger put in Spandau Prison and then "suicided"? Did the British use an early form of mind control on Hess' double? John Foster Dulles of the OSS and CIA suspected as much. Farrell also uncovers the strange death of Admiral Richard Byrd's son in 1988, about the same time of the death of Hess.

288 Pages. 6x9 Paperback. Illustrated. $19.95. Code: HAPG

HIDDEN FINANCE, ROGUE NETWORKS & SECRET SORCERY
The Fascist International, 9/11, & Penetrated Operations
By Joseph P. Farrell

Farrell investigates the theory that there were not *two* levels to the 9/11 event, but *three*. He says that the twin towers were downed by the force of an exotic energy weapon, one similar to the Tesla energy weapon suggested by Dr. Judy Wood, and ties together the tangled web of missing money, secret technology and involvement of portions of the Saudi royal family. Farrell unravels the many layers behind the 9-11 attack, layers that include the Deutschebank, the Bush family, the German industrialist Carl Duisberg, Saudi Arabian princes and the energy weapons developed by Tesla before WWII.

296 Pages. 6x9 Paperback. Illustrated. $19.95. Code: HFRN

THRICE GREAT HERMETICA & THE JANUS AGE
By Joseph P. Farrell

What do the Fourth Crusade, the exploration of the New World, secret excavations of the Holy Land, and the pontificate of Innocent the Third all have in common? Answer: Venice and the Templars. What do they have in common with Jesus, Gottfried Leibniz, Sir Isaac Newton, Rene Descartes, and the Earl of Oxford? Answer: Egypt and a body of doctrine known as Hermeticism. The hidden role of Venice and Hermeticism reached far and wide, into the plays of Shakespeare (a.k.a. Edward DeVere, Earl of Oxford), into the quest of the three great mathematicians of the Early Enlightenment for a lost form of analysis, and back into the end of the classical era, to little known Egyptian influences at work during the time of Jesus.

354 Pages. 6x9 Paperback. Illustrated. $19.95. Code: TGHJ

ROBOT ZOMBIES
Transhumanism and the Robot Revolution
By Xaviant Haze and Estrella Eguino,

Technology is growing exponentially and the moment when it merges with the human mind, called "The Singularity," is visible in our imminent future. Science and technology are pushing forward, transforming life as we know it—perhaps even giving humans a shot at immortality. Who will benefit from this? This book examines the history and future of robotics, artificial intelligence, zombies and a Transhumanist utopia/dystopia integrating man with machine. Chapters include: Love, Sex and Compassion—Android Style; Humans Aren't Working Like They Used To; Skynet Rises; Blueprints for Transhumans; Kurzweil's Quest; Nanotech Dreams; Zombies Among Us; Cyborgs (Cylons) in Space; Awakening the Human; more. Color Section.

180 Pages. 6x9 Paperback. Illustrated. $16.95. Code: RBTZ

ANCIENT ALIENS & SECRET SOCIETIES
By Mike Bara
Did ancient "visitors"—of extraterrestrial origin—come to Earth long, long ago and fashion man in their own image? Were the science and secrets that they taught the ancients intended to be a guide for all humanity to the present era? Bara establishes the reality of the catastrophe that jolted the human race, and traces the history of secret societies from the priesthood of Amun in Egypt to the Templars in Jerusalem and the Scottish Rite Freemasons. Bara also reveals the true origins of NASA and exposes the bizarre triad of secret societies in control of that agency since its inception. Chapters include: Out of the Ashes; From the Sky Down; Ancient Aliens?; The Dawn of the Secret Societies; The Fractures of Time; Into the 20th Century; The Wink of an Eye; more.
288 Pages. 6x9 Paperback. Illustrated. $19.95. Code: AASS

THE CRYSTAL SKULLS
Astonishing Portals to Man's Past
by David Hatcher Childress and Stephen S. Mehler
Childress introduces the technology and lore of crystals, and then plunges into the turbulent times of the Mexican Revolution form the backdrop for the rollicking adventures of Ambrose Bierce, the renowned journalist who went missing in the jungles in 1913, and F.A. Mitchell-Hedges, the notorious adventurer who emerged from the jungles with the most famous of the crystal skulls. Mehler shares his extensive knowledge of and experience with crystal skulls. Having been involved in the field since the 1980s, he has personally examined many of the most influential skulls, and has worked with the leaders in crystal skull research, including the inimitable Nick Nocerino, who developed a meticulous methodology for the purpose of examining the skulls.
294 pages. 6x9 Paperback. Illustrated. Bibliography. $18.95. Code: CRSK

AXIS OF THE WORLD
The Search for the Oldest American Civilization
by Igor Witkowski
Polish author Witkowski's research reveals remnants of a high civilization that was able to exert its influence on almost the entire planet, and did so with full consciousness. Sites around South America show that this was not just one of the places influenced by this culture, but a place where they built their crowning achievements. Easter Island, in the southeastern Pacific, constitutes one of them. The Rongo-Rongo language that developed there points westward to the Indus Valley. Taken together, the facts presented by Witkowski provide a fresh, new proof that an antediluvian, great civilization flourished several millennia ago.
220 pages. 6x9 Paperback. Illustrated. References. $18.95. Code: AXOW

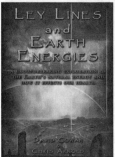

LEY LINE & EARTH ENERGIES
An Extraordinary Journey into the Earth's
Natural Energy System
by David Cowan & Chris Arnold
The mysterious standing stones, burial grounds and stone circles that lace Europe, the British Isles and other areas have intrigued scientists, writers, artists and travellers through the centuries. How do ley lines work? How did our ancestors use Earth energy to map their sacred sites and burial grounds? How do ghosts and poltergeists interact with Earth energy? How can Earth spirals and black spots affect our health? This exploration shows how natural forces affect our behavior, how they can be used to enhance our health and well being.
368 PAGES. 6x9 PAPERBACK. ILLUSTRATED. $18.95. CODE: LLEE

SAUCERS, SWASTIKAS AND PSYOPS
A History of a Breakaway Civilization
By Joseph P. Farrell

Farrell discusses SS Commando Otto Skorzeny; George Adamski; the alleged Hannebu and Vril craft of the Third Reich; The Strange Case of Dr. Hermann Oberth; Nazis in the US and their connections to "UFO contactees"; The Memes—an idea or behavior spread from person to person within a culture—are Implants. Chapters include: The Nov. 20, 1952 Contact: The Memes are Implants; The Interplanetary Federation of Brotherhood; Adamski's Technological Descriptions and Another ET Message: The Danger of Weaponized Gravity; Adamski's Retro-Looking Saucers, and the Nazi Saucer Myth; Dr. Oberth's 1968 Statements on UFOs and Extraterrestrials; more.

272 Pages. 6x9 Paperback. Illustrated. $19.95. Code: SSPY

LBJ AND THE CONSPIRACY TO KILL KENNEDY
By Joseph P. Farrell

Farrell says that a coalescence of interests in the military industrial complex, the CIA, and Lyndon Baines Johnson's powerful and corrupt political machine in Texas led to the events culminating in the assassination of JFK. Chapters include: Oswald, the FBI, and the CIA: Hoover's Concern of a Second Oswald; Oswald and the Anti-Castro Cubans; The Mafia; Hoover, Johnson, and the Mob; The FBI, the Secret Service, Hoover, and Johnson; The CIA and "Murder Incorporated"; Ruby's Bizarre Behavior; The French Connection and Permindex; Big Oil; The Dead Witnesses: Guy Bannister, Jr., Mary Pinchot Meyer, Rose Cheramie, Dorothy Killgallen, Congressman Hale Boggs; LBJ and the Planning of the Texas Trip; LBJ: A Study in Character, Connections, and Cabals; LBJ and the Aftermath: Accessory After the Fact; The Requirements of Coups D'État; more.

342 Pages. 6x9 Paperback. $19.95 Code: LCKK

THE TESLA PAPERS
Nikola Tesla on Free Energy &
Wireless Transmission of Power
by Nikola Tesla, edited by David Hatcher Childress

David Hatcher Childress takes us into the incredible world of Nikola Tesla and his amazing inventions. Tesla's fantastic vision of the future, including wireless power, anti-gravity, free energy and highly advanced solar power. Also included are some of the papers, patents and material collected on Tesla at the Colorado Springs Tesla Symposiums, including papers on: •The Secret History of Wireless Transmission •Tesla and the Magnifying Transmitter •Design and Construction of a Half-Wave Tesla Coil •Electrostatics: A Key to Free Energy •Progress in Zero-Point Energy Research •Electromagnetic Energy from Antennas to Atoms

325 PAGES. 8x10 PAPERBACK. ILLUSTRATED. $16.95. CODE: TTP

COVERT WARS & THE CLASH OF CIVILIZATIONS
UFOs, Oligarchs and Space Secrecy
By Joseph P. Farrell

Farrell's customary meticulous research and sharp analysis blow the lid off of a worldwide web of nefarious financial and technological control that very few people even suspect exists. He elaborates on the advanced technology that they took with them at the "end" of World War II and shows how the breakaway civilizations have created a huge system of hidden finance with the involvement of various banks and financial institutions around the world. He investigates the current space secrecy that involves UFOs, suppressed technologies and the hidden oligarchs who control planet earth for their own gain and profit.

358 Pages. 6x9 Paperback. Illustrated. $19.95. Code: CWCC

BIGFOOT NATION
A History of Sasquatch in North America
By David Hatcher Childress

Childress takes a deep look at Bigfoot Nation—the real world of bigfoot around us in the United States and Canada. Whether real or imagined, that bigfoot has made his way into the American psyche cannot be denied. He appears in television commercials, movies, and on roadside billboards. Bigfoot is everywhere, with actors portraying him in variously believable performances and it has become the popular notion that bigfoot is both dangerous and horny. Indeed, bigfoot is out there stalking lovers' lanes and is even more lonely than those frightened teenagers that he sometimes interrupts. Bigfoot, tall and strong as he is, makes a poor leading man in the movies with his awkward personality and typically anti-social behavior. Includes 16-pages of color photos that document Bigfoot Nation!

320 Pages. 6x9 Paperback. Illustrated. $22.00. Code: BGN

MEN & GODS IN MONGOLIA
by Henning Haslund

Haslund takes us to the lost city of Karakota in the Gobi desert. We meet the Bodgo Gegen, a god-king in Mongolia similar to the Dalai Lama of Tibet. We meet Dambin Jansang, the dreaded warlord of the "Black Gobi." Haslund and companions journey across the Gobi desert by camel caravan; are kidnapped and held for ransom; witness initiation into Shamanic societies; meet reincarnated warlords; and experience the violent birth of "modern" Mongolia.

358 Pages. 6x9 Paperback. Illustrated. $18.95. Code: MGM

PROJECT MK-ULTRA
AND MIND CONTROL TECHNOLOGY
By Axel Balthazar

This book is a compilation of the government's documentation on MK-Ultra, the CIA's mind control experimentation on unwitting human subjects, as well as over 150 patents pertaining to artificial telepathy (voice-to-skull technology), behavior modification through radio frequencies, directed energy weapons, electronic monitoring, implantable nanotechnology, brain wave manipulation, nervous system manipulation, neuroweapons, psychological warfare, satellite terrorism, subliminal messaging, and more. A must-have reference guide for targeted individuals and anyone interested in the subject of mind control technology.

384 pages. 7x10 Paperback. Illustrated. $19.95. Code: PMK

LIQUID CONSPIRACY 2:
The CIA, MI6 & Big Pharma's War on Psychedelics
By Xaviant Haze

Underground author Xaviant Haze looks into the CIA and its use of LSD as a mind control drug; at one point every CIA officer had to take the drug and endure mind control tests and interrogations to see if the drug worked as a "truth serum." Chapters include: The Pioneers of Psychedelia; The United Kingdom Mellows Out: The MI5, MDMA and LSD; Taking it to the Streets: LSD becomes Acid; Great Works of Art Inspired and Influenced by Acid; Scapolamine: The CIA's Ultimate Truth Serum; Mind Control, the Death of Music and the Meltdown of the Masses; Big Pharma's War on Psychedelics; The Healing Powers of Psychedelic Medicine; tons more.

240 pages. 6x9 Paperback. Illustrated. $19.95. Code: LQC2

ORDER FORM

10% Discount When You Order 3 or More Items!

One Adventure Place
P.O. Box 74
Kempton, Illinois 60946
United States of America
Tel.: 815-253-6390 • Fax: 815-253-6300
Email: auphq@frontiernet.net
http://www.adventuresunlimitedpress.com

ORDERING INSTRUCTIONS

✓ Remit by USD$ Check, Money Order or Credit Card

✓ Visa, Master Card, Discover & AmEx Accepted

✓ Paypal Payments Can Be Made To:
  info@wexclub.com

✓ Prices May Change Without Notice

✓ 10% Discount for 3 or More Items

SHIPPING CHARGES

United States

✓ Postal Book Rate { $4.50 First Item / 50¢ Each Additional Item

✓ POSTAL BOOK RATE Cannot Be Tracked!
  Not responsible for non-delivery.

✓ Priority Mail { $6.00 First Item / $2.00 Each Additional Item

✓ UPS { $7.00 First Item / $1.50 Each Additional Item

NOTE: UPS Delivery Available to Mainland USA Only

Canada

✓ Postal Air Mail { $15.00 First Item / $2.50 Each Additional Item

✓ Personal Checks or Bank Drafts MUST BE
  US$ and Drawn on a US Bank

✓ Canadian Postal Money Orders OK

✓ Payment MUST BE US$

All Other Countries

✓ Sorry, No Surface Delivery!

✓ Postal Air Mail { $19.00 First Item / $6.00 Each Additional Item

✓ Checks and Money Orders MUST BE US$
  and Drawn on a US Bank or branch.

✓ Paypal Payments Can Be Made in US$ To:
  info@wexclub.com

SPECIAL NOTES

✓ RETAILERS: Standard Discounts Available

✓ BACKORDERS: We Backorder all Out-of-
  Stock Items Unless Otherwise Requested

✓ PRO FORMA INVOICES: Available on Request

✓ DVD Return Policy: Replace defective DVDs only

ORDER ONLINE AT: www.adventuresunlimitedpress.com

10% Discount When You Order 3 or More Items!

Please check: ✓

| ☐ This is my first order | ☐ I have ordered before |

Name
Address
City
State/Province — Postal Code
Country
Phone: Day — Evening
Fax — Email

Item Code	Item Description	Qty	Total

Please check: ✓

	Subtotal ▶	
Less Discount-10% for 3 or more items ▶	
☐ Postal-Surface	Balance ▶	
☐ Postal-Air Mail (Priority in USA)	Illinois Residents 6.25% Sales Tax ▶	
	Previous Credit ▶	
☐ UPS (Mainland USA only)	Shipping ▶	
	Total (check/MO in USD$ only) ▶	
☐ Visa/MasterCard/Discover/American Express		

Card Number:

Expiration Date: — Security Code:

✓ SEND A CATALOG TO A FRIEND: